INTERNATIONAL DESIGN YEARBOOK 18

Based on an original idea by Stuart Durant

Senior Editor: Cleia Smith
Project Manager: Ros Gray

First published in the United States in 2003 by Abbeville Press
116 West 23rd Street
New York, NY 10011

First published in Great Britain in 2003 by Laurence King Publishing
71 Great Russell Street
London WC1B 3BP

Printed in China

First Edition

ISBN 0-7892-0788-5

10 9 8 7 6 5 4 3 2 1

Library of Congress Cataloging-in-Publication Data available upon request

'Design is the whole experience of living'. Karim Rashid

INTERNATIONAL DESIGN YEARBOOK 18

General Editor: Jennifer Hudson, Design Director: Karim Rashid, Design: Nomad-ic

'Design is the whole experience of living'. Karim Rashid

Edited by Karim Rashid
Abbeville Press Publishers
New York London

GLOBAL DESIGN

Karim Rashid

As the twenty-first century progresses, a global design phenomenon is becoming apparent. I hesitate to call it a movement since it is too broad, eclectic and indefinable, yet there are several contiguous design languages being articulated. We live in a time of no real 'school of thought' or axiom of design. There is no prevailing dogma, no real rules; as Rem Koolhaas would say, no 'ball and chains'. In a sense there is a lack of direction, but paradoxically and quite fortunately the dropping of 'design protocols' is more freely allowing a plethora of dynamic explorations, ideas and styles to coexist in an entropic worldwide lifestyle-condition.

'Worldwide' is a heuristic concept – a reduction of signs to comprehend large complex phenomena. The 'welt' of our human existence means the relationship of mankind with the world, or time with man. 'World' is thus equivalent to time, and 'wide' implies the limitless possibilities of progress, evolution and a spirit of borderless creation. Since we cannot define the world in our lifetime, we define time by rigorously pursuing new possibilities and original thought. This is what defines time and in turn our lifeworld, in as much as we can comprehend it. I believe that 'local' is not relative anymore; it implies a way of operating without thinking, a myopic way of seeing human behaviour in a narrow context and working within a cultural and environmental vacuum. The world is shrinking, and 'global' is the only way to perceive culture and think freely, unobstructed by political, social, historical, hierarchical, nepotistic and class issues. Local thinking is the nemesis of free spirit, and real creative thought in our new condition of globalization questions the sanctity of the historic built environment. With somewhat flippant oversimplification, I want to set the stage for the twenty-first century and introduce the themes that structure this book:

We know that the pure clean geometry of the machine age and Bauhaus was as much a style as it was a dialectic of production methods. We have witnessed the over-stylized streamlined 1930s, the organic movement and nature movement of the 1940s (and the postwar frugal design trend), the decadence of the popstractive and kinetic 1950s, the space-age optimistic 1960s, the decadent gloss and chrome of the orgiastic 1970s, the acrimonious postmodern electropop 1980s, the indecisiveness and eclecticism of the 1990s, and now what I will call the global 'culturally unidentifiable' typology or morphology of the 2000s.

The 'zeros', a time when commodity is more poetic through personalization than movement, has several directions. The inspirations of the past are extremely evident – objects and space are borrowed freely and without guilt in a hyper-retro, neo-postmodern spirit. I call it FUTURETRO. There is an especially strong simulation of the 1960s – a time of futuristic optimism, of a human desire to go beyond the Earth, and to reach a higher technological state of humanity. Once we stepped on the moon in 1969 it all ended; we realized that the dead rock has no future. In turn we focused back on our own planet, some of us became ecologically concerned, while others retaliated with 'no future', the rebellious 'design is dead' or, as Theodor Adorno stated, 'the end of art'. So we had a 'granola movement' that by the end of the 1970s became artificial and hollow (yet wildly corybantic, slick and snow white). Then we grew into the conspicuous consumption of the yuppie 1980s, and ended up in the modernist revival doctrine of 1990s minimalism. This manifested itself as a style rather than the intellectual mantra 'less is more'. We cannot call that tendency 'pure' because pure refers to a spiritual and utopian state of nothingness, where one reaches the ultimate solution for a given problem, as in many of the examples of MINIMAL designs illustrated here.

Conversely, designers seem increasingly to be dropping the elitism of 'taste' that pinnacled in the 1990s and to be realizing the subversive messages of eclecticism and NUKITSCH. I see a strong revival of this ever-present typology as it has re-emerged in a more witty or 'designed' commentary of humour and sarcasm. Kitsch is of new interest because it asks questions, and opposes radical but literal proposals for our 'buyosphere' (a term coined by Thomas Hine). The past and present are simultaneously superimposed, and everything and anything that existed in the past is attainable, achievable, accessible and perversely acceptable.

Many proposals submitted for this year's 'International Design Yearbook' illustrate one of my most sacred interests: the PHENOMENA of objects, or phenomenological objects that shift and change over time and space. This involves the heightened unveiling of a surprise, a phenomenological condition where the event is never repeated, but always different. New 'smart' materials and technologies are offering us a more dynamic, visual and illusory object culture.

The other more contrived form of such phenomenological objects is MULTIPLICITY. We just can't help ourselves – we love the idea of creating one more function in a product (what corporate product designers call 'features'), or creating reconfigurable, variable paradigms for interaction, personalization and experience. Choice and variety come from flexible products. The opportunity to change something increases our awareness of our own time and being. Moreover, surface change, flexibility and customization have become an extension of our desktop publishing world, where EMBELLISHMENT is a way of signifying identity and expressing individuality.

I have a strong affinity towards amorphous form, as it is extremely human yet challenging and complex in its data. In recent years I have witnessed an ORGANIC blobular morphology of our land-scape. I call this amorphous movement Blobism or Organicism. Human sensibility here combines with digital tools to morph and create complex geometry that never before was definable and retainable mathematically – a direct relationship with chaos, fuzzy logic, parametric computing and animation.

I believe we are entering into an era of TECHNOcracy – the democracy of technology – in which technology inspires and enlightens physical and virtual fields of design worldwide. As technology autonomizes our playing field and information instantaneously touches all parts of the globe, it often seems as though we are all reading the same books, watching the

Opposite: Dish soap. Method, USA
Above: Butterfly chair. Injection-moulded ABS, chromed steel. Magis srl, Italy

same movies, looking at the same magazines, listening to the same music, eating the same food, even drinking the same soda. As with money and language, inevitably our creative manifestations are starting to converge. Our new global language is binary notation – zeros and ones. This new language is broader, more complex and accurate than any previously developed by the human race. The concept of simultaneous creative discovery is based on shared information and so, in turn, we are shaping similar ideas and constructs, and shaping similar notions of physical space and time. Yet the fact that at the same time there is no single dogma like that of Modernism fifty years ago means that humans can exist and domesticate with a new freer spirit.

The results are everything, everyone, every sensibility and every idea appearing simultaneously in real-time anywhere and every-where. I love the autonomy of an eclectic landscape, where all human creative thought may be expressed regardless of origins or history. We live in a time of NO RULES and NO BORDERS, and of real freedom of expression. I like this contradiction, where we are influenced by everything, and where all information moves simulta-neously. Things may appear homogenous globally, but at the same time they are split into micro-tribes, into thousands of diverse sensibilities, of multiple tastes, views, beliefs, values and feelings. These contradicting scenarios of the global village versus an eclectic chaos in actuality form our new contemporary aesthetic world – a plethora of expressions, signs and logos, where there is no good and bad, like the Nietzschian notion of 'beyond good and evil'.

Modernity is an old concept, a way of creating a scientific religion that is caught within the strictures of Cartesian thinking. The twenty-first century is about a new energy, a material–immaterial marriage of formlessness and form, of transparency and colour, a kinesthetic binary hypercontextual existence, a digital nature, a techno-organic world, a kaleidoscope of diverse yet elevated experiences. The digital age represents this place – real and virtual, metaphysical and physical spaces layered together. Objects smell, taste, breathe, touch and participate in our experiences.

Companies will market individualization to address smaller and smaller 'specialized' groups, as the Internet already demonstrates today. A car will be completely customized down to the body form, as will be fragrances, running shoes, even body parts. Today this trend of Variance™ is seen in customized laser-cut Levi's jeans, bicycle manufacturing, computer configurations, and so on. Manufacturers will utilize new 4-D computer numeric machinery and other sophisticated methods in mass production cycles to create one-off individually specified designs. The other scenario is consumers using visual programs on the Internet to morph, vary and personalize products. This 'tool path' information would be digitally transferred at the manufacturer's site and then produced and delivered to the individual.

Art embraces contemporary myth with contradictory results. Design also celebrates myth, but solicits less value. As soon as the art object shifts into the role of utility, or implies a functional role, the value shifts to 'things of everyday life'. Why should such com-modities carry any real value as cultural artifact or manifestation of preconsciousness? The ubiquity of produced goods does not necessarily devalue their presence. As Duchamp's ready-mades shifted our view of the banal object and questioned what art might be, so mass-produced consumer goods today – from Warhol wallpaper, Cocteau dishes, Munch's 'Scream' as an inflatable doll to T-shirts and screensavers with Van Gogh on them – shift our view of art from sacred to kitsch. Disposability, change and new technologies all speak of a vast inertia of consumption, of less innate value. Eames chairs, although considered critically important to American design history, have distant perceived value, yet millions were produced, hence why there are so many in circulation. It is equivalent to finding 'Saturday Night Fever' vinyl LPs in thrift stores.

So how is it possible that design can infiltrate the context of 'high' culture? And can culture exist within the realm of hyper-consumption, of multitudes of new models, as the machine of fashion, trend and consumption? Should it? What is art's contribution, and how does art distinguish itself from the plethora of images we see every day in the media? Is real art now part of the commercial world, where it pushes greater boundaries and higher aesthetics, and forges ever greater experimentation at a belligerent pace?

I see the future of our aesthetic world crossing all disciplines, so that design, art, architecture, fashion and music fuse together to increase our experiences and bring greater pleasure to our material and immaterial lives. As production technologies near perfection we can create incredibly sophisticated goods. The consumer is becoming very savvy, and has higher expectations from the built environment and from their experiences. Experiences need to be increased, pleasure needs to be addressed, and design is a tool to reshape our phenomenological and evanescent public memory. Our motivations should focus around a desire to fill our collective conscious with ideas that are seamless connections between art and life, between dreams and reality. As art takes its ideas from everyday life, I hope that everyday life will take its ideas from art. We live in two diametric worlds: one moving towards universal sameness and nostalgia, and the other towards individualized freedom of expression, choice and creativity.

I hope for world peace and world love. I send my regards and prayers to my friends, my colleagues and everyone who was touched by the tragedy of September 11th. I have lost some friends in this catastrophe and it hurts, but more importantly I am concerned about us as a race, and our civility. We need to start loving each other and forge ahead to create beauty for everyone.

'Be the change you want to see in the world' (Mahatma Gandhi)

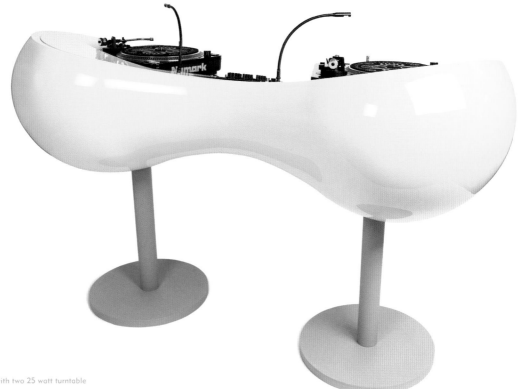

DJ Kreemy Table. Moulded fiberglass body and steel base with two 25 watt turntable lamps and internal power bar. Pure Design, Canada

In 2002 'The International Design Yearbook' took the unprecedented step of changing its format. As design grows in importance both economically and socially it is no longer interesting enough to present a catalogue of what the preceding year had to offer, and to divide that selection into recognizable but inevitably rather mundane categories. To a more discerning public there is a need for a thesis – to bring the typologies into a dialogue with the overall ideology of the guest editor.

Last year Ross Lovegrove organized his selection into materials, which reflects his preoccupation with experimenting with new substances or using established ones in an adaptive way in order to create his signature designs. His 'Organic Essentialism' takes references from the natural world and ergonomic moulding forms, which reduce the elements to the bare minimum and require the use of materials that bend and morph, and which are light and environmentally friendly. His philosophy, of which the above is but a small part, is personal and convincing. It set a hard act to follow, making our choice of the following guest editor a difficult one.

Karim Rashid was the obvious contender. With the enormously successful publication of his monograph 'I Want to Change the World', which sold 15,000 copies worldwide, he was clearly someone with a lot to say and who the market wanted to listen to. Karim was born in Cairo of Egyptian and English parentage. He lived most of his life in Canada but now works in New York. He thus also answered to our desire to find a designer strongly associated with America to offer his or her critique on what was currently important to the design community. Europeans generally look on the United States as being without style. Despite having been a world leader in the post-war period, with designers such as Raymond Loewy, Charles and Ray Eames, Eero Saarinen and Florence Knoll, it seems to have lost its way during the following decades. It is only now re-emerging thanks to designers such as Karim, who is on a mission to bring good quality design to the shopping malls of Middle America.

Karim describes himself as an intellectual and a cultural editor, and sees the ideas behind his work, or rather the importance he places on design as an instigator of social change, as in some ways more important than the products he creates. He is a man of missionary zeal and utopian ideals who wants as many people as possible to hear his message, and for that message to change the way we perceive the world. Even his mobile phone voices the words of Gandhi – 'Be the change you want to see in the world'.

One of Karim's greatest inspirations is our move into the twenty-first century – the 'zeros' as he calls it. New cultures demand new forms, materials and styles, and above all he wants the items, interiors and graphics that he creates to reflect the new digital age, which he sees as the third industrial revolution. He proposes new objects for new behaviours and produces work that suits the

Sensual Minimalism

Stackable chair, Oh. Injection-moulded polypropylene, powder-coated steel, Umbra, USA modern, cosmopolitan, casual, trendy, laid-back consumer of the computer society. However, for him design does not just serve or reflect changes in the way we live but is instrumental, as it is capable of altering human behaviour and creating new social conditions. As Karim explains, 'Design is where we define and shift physical space, where we affect physical, psychological and sociological behaviour and set up conditions of human experience.' We are experiencing a unique evolution in the way we live and the items that surround us should mirror that progress. Form should not follow function, rather it should follow subject, that subject being not necessarily the object's function but the social and philosophical reason why it exists at a given time.

His didacticism and strong opinions have earned Karim the reputation of being arrogant, but this is a misplaced criticism, as is obvious to anyone who has worked closely with him. He is incredibly prolific, having created over 800 designs, many of which have earned much press coverage due to their ubiquitous and plural nature. This has led to a certain amount of professional jealousy. Karim considers the four important requirements for success to be rigour, consistency, perseverance and talent. He is enthusiastic to the point of obsessive-ness, and that enthusiasm communicates itself – he thinks aloud and is constantly besieged by ideas. Within ten minutes of sitting down with him to discuss the Yearbook he had not only decided on its format, layout and premise, but had also come up with a couple of thoughts about its cover. The fifteen staff at his design studio in the Chelsea district of Manhattan all share his vision and there is a constant cross-fertilization of ideas between them and their employer. When I met him in New York, we were repeatedly joined by members of his team bringing him different projects to discuss, and he was as prepared to embrace their concepts as to impart his own.

Karim refers to a new phenomenon for a new century, one that will be global in scope. The future of design will be 'customization, personalization, robotic production, smart technology, new polymers, new behaviours, organic geometry, sensualism, new production methods, experiential, phenomenological, non-serialised, soft, friendly, borderless, seamless and all-encompassing'. It is a design approach that should be accessible to all. He is the moving force behind the democratic design wave, the 'Garbo Can' being the most famous of his elegant, sculptural household objects that are sold at reasonable prices. Design should thus be disposable and biodegradable, with a built-in obsolescence of five years so that it is forced to continually re-invent itself and adapt to different social conditions as the century progresses. His aim is not only to produce objects that are available to all, but also to demystify design by making it more mainstream. 'Design must become more visible in a world of media so that design is of pedestrian interest and desire rather than a marginal subject'. Above all, Karim is concerned with the consumer. A design should be de-stressing; it should add to the quality of life; if it does not then there is no place for it. It should be enjoyable, make life easier and appeal to our notions of beauty, as well as addressing specific problems. When designing the 'Oh' chair for Umbra, he changed its measurements to make the seat three inches wider 'after seeing the guys in the factories with their big asses'. 'European chairs are too delicate. America needs bigger, stronger, more casual furniture'.

Karim has coined the expression 'Sensual Minimalism' to describe his style, which has an emotional appeal while staying minimal. Soft and tactile forms are more human, signifying comfort and pleasure. Our response to them is from within; it is a primaeval memory of the breast and of nurturing. The blob is Karim's trademark. For him it is both the physical manifestation of this philosophy and a visual reaction to mechanization. It is ever-changing, non-serial and is an expression of the biological nature of all existence. His 'Blobject' was originally conceived as a one-off for the Sandra Gering Gallery in New York, made in automotive lacquer on fibreglass. It has gone through several metamorphoses, being produced as a limited batch production for Trans>6, this time in vinyl-dipped soft PVC, and it is now part of the 2002 Edra collection.

Unlike Ross Lovegrove, Karim's organic style does not spring from observing natural artefacts. Rather it stems from the technical advances achieved by new computer software, the shapes that have been liberated, and the need for materials that can be moulded into these complicated soft forms. He creates natural configurations from digital processes. Karim has spoken of a computer programme called Metaball that allows different geometric shapes to be blended, creating a 'plastic, biomorphic shape that can be regulated'. This can then be cast in materials that are mutable and flexible, and which have new tactile surfaces – synthetic rubbers, silicones, santoprenes, double injection-moulded polymers. The result is an object that is sensuous to the eye as well as experiential.

Karim is convinced that future experimentation lies in 'smart' materials, which will have the capacity to interact with us. His 7.6 metre-wide sofa 'Momo Pink 100' was first shown at the New York gallery Deitch Projects, but his desire is eventually to cover it in 'Smartwear' so that the user can audio-video telecommunicate while relaxing. Ironically, technology has also liberated design from the uniformity it originally imposed upon it. Increased mechanization initially resulted in the destruction of individuality. Karim strongly believes that customization and variance of mass-produced objects is the only way to reconcile the machine and the human. CAD programs and smart materials have enabled idiosyncratic yet cost-effective manufacturing. His crystal 'Morph' vase for the American company Nambé is a mass-produced yet non-serialized object. The cut lines were designed with a digital tool path so that they change randomly from vase to vase. The cuts are forever different due to the random software program.

The greatest democratization of design, however, will come when anybody can create their own designs. Advanced computer technology is well on its way to enabling this. Karim has come up with the idea of 'desk-top manufacturing', where the consumer will be able to build up a three-dimensional object using such devices as a 3-D printer, or alternatively personalize a product by engaging with a visual program on the Internet. These designs will then be transferred to the manufacturer's site, produced and delivered. Karim is not concerned that this would lead to a dumbing down, as cream always rises. 'We can all taste, but still somebody somehow will manage to be a connoisseur about food where the rest are not.' He believes the individual should be allowed to change his own environment, and yet there will always be a place for the expert with talent. The professional market should adapt accordingly: 'Less is more'. Karim would like to convince manufacturers to streamline their output, producing only a handful of truly remarkable objects each year.

As a mediator between industry and the user, an 'artist of real issues', Karim's work is a perfect blend between professionally executed products, which result from thorough understanding and appreciation of production methods, and today's more personalized, communicative and interactive design. Karim promised himself that when he had a practice he would not become a 'service' to clients, but instead would offer them his vision and sensibility. He is aware that his clients are not artistic patrons and as such he needs to consider their preoccupation with profit. Yet his very personal style is not something he will compromise, and he sees it as the role of the designer to intuit how people live, behave and change. It is this symbiosis that results in a successful commodity without the sacrifice of artistic creativity. For many years he told me that he played it safe working for conservative clients who would not authorise anything that differed from their predictable expectations. Nambé were the first to recognize his forward-looking potential and give him the freedom to pursue his beliefs. Since then he has chosen only those manufacturers insightful enough to mix industry with conviction.

Karim was raised in an artistic family as his father is a painter and set designer. It was his father who taught him to observe and record (at the age of four he could draw in perfect perspective), and who instilled in him the belief that he was capable of designing anything and of touching all aspects of his physical landscape. Every weekend Karim's father would take him to the television design studios and there Karim would watch, learn and copy. The family home was full of monographs on artistic pluralists such as Andy Warhol, Rodchenko, Picasso, Corbusier, Buckminster Fuller, Eames, Yves Saint Laurent and Raymond Loewy, all of whom he admired and who have influenced his career.

His background informed the way he looks at the relationship between art and design. He considers that in every profession there are artists who are distinguishable. These few are truly aware of their surroundings, and see messages, ideas, signs and the broader issues; they are passionate about creating something original. Like an artist in front of the life model, Karim learned his profession and saturated his knowledge. Once he had achieved this aim he set out to forget all he had been taught as only with this cleansing process would his mind find the freedom to act autonomously and to create. When asked whether he judged some of his objects to be pieces of art, he replied that he is equally interested in designing a banal product sold for $1 as he is in creating one-off pieces that exist only in art galleries. He presently shows at three exhibition spaces in New York.

We live in a borderless world where all creative disciplines are blurring, merging and hybridizing. It is Karim's aim to form a Warholian factory that produces buildings, interiors, fashion, objects, products, furniture, art, installations, music and film. Among the forty-five projects his studio is currently working on, there are interiors of hotels in Brighton and London, the architecture and interior of another in Athens, a night-club in New York, jewellery for a Swiss company, a new plastic chair, a clothing collection, a line of high-tech products and a radical furniture system. Karim is also cutting his first album, which followed on from the success of his six-minute dance track. 'Plob' was an installation-cum-interactive musical experience staged at the Capp Street Gallery in San Francisco. 'Plob' is a noun or verb that was invented by Karim. It is the stage between liquid plastic and solid material and also a state of being. For the installation, the gallery was turned into an amorphous plastic scape denoting a world without boundaries. The white mouldings were rearrangeable abstractions, or 'plobs', surrounded by a field of fluorescence, which alluded to the world of technology. Motion sensors inside each 'plob' were linked to tracks of music Karim had composed and to coloured lights. Moving through the installation caused different combinations to be activated, the more people participating the more sensors were triggered and the more complex the music became.

Karim believes that the objects that shape our lives should be 'trans-conceptual, multicultural hybrids'. As he mentions in his introduction, 'we live in a time of no real 'school of thought' or axiom of design ... no prevailing dogma, no real rules'. His selection is categorized to reflect this. Each chapter deals with a theme that he feels is currently important – Futuretro, Nukitsch, Phenomena, Embellishment, Organic, Multiplicity, Minimum and Techno – a need for nostalgia and historical reference, combined with a reaction to the visual standardization of minimalism and an exploration of the technological advances of the digital age. He has also chosen pieces that correspond to his personal priorities; from the democratic designs of Philippe Starck's Target lines, to the variance of Gaetano Pesce's 'Nobody's Perfect' range; from the casual aesthetic of Jerszy Seymour's 'Muff Daddy' or the Campana brothers' 'Boa' to the computer-generated designs of Nathelie Jean and the 'Aspetto' chair by Catherine Lorenz and Stefano Kaz. His love of innovative materials is evident in his selection of Tokujin Yoshioka's 'Honey-Pop' chair and the photo-reactive substances that can be found in the Phenomena section. His belief that good design can create new social conditions is highlighted in the inclusion of 'Sparrow', an electric car designed for one person, which suggests a change in the way we conceive city transportation.

In 'I Want to Change the World', the musician David Byrne writes that since the mid-1990s there has been a growing design current that emphasizes joy, exuberance and optimism, all of which are summed up in the work of Karim Rashid. Design does not stand on its own but is part of our sociological, psychological and political world – 'design is the whole experience of living'.

Crystal from assorted collections, Nambé, USA

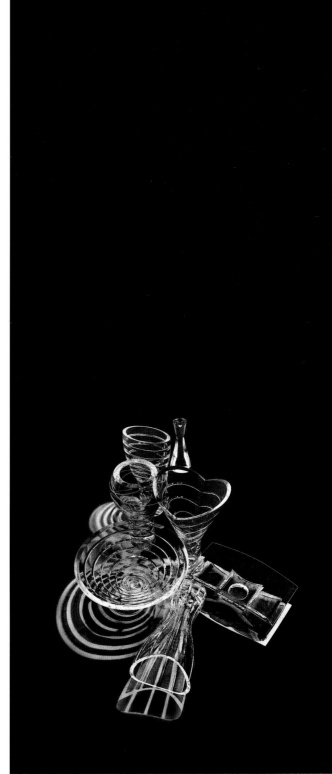

In this new century we are witnessing an exhaustive excavation into the vaults of the past, even more so than the post-modern 1980s. It seems as if history is unfolding at a hypertrophic rate as we search into the past to unveil everything and anything, be it good or bad. Is it that the yuppies and boomers currently empowered in the market are trying to revisit their past, and are not willing to let go? Is it a fear of mortality? Maybe we collectively see the past as more interesting than now and so we copy it without compunction, consciously or subconsciously. Possibly because of similar needs exhibited in past human behaviour, we also pretend the past did not exist. This is a phenomenon of a 'generation of the now'. The difference of this revival in comparison to previous neo-movements is the new technologies that we can bring to the real / virtual revisitation. The retrograde movement of the 1960s seems to be the most inspiring because at that time there was a strong futurist optimism, a belief that we would reach another universe, that we would live in a seamless techno-organic world, and that there must be more to our existence than the planet Earth. Now we can remake the past better. If you redo a chair today that existed in the 1960s you can greatly improve the entire process and end result thanks to technological advances. But let's admit that it is easier to simulate the past than predict the future...

'To articulate the past historically does not mean to recognize it "the way it really was" ...
It means to seize hold of a memory as it flashes up at a moment of danger.' Walter Benjamin

Previous page: Karim Rashid. Armchair, Swivel. Upholstered seat with removable fabric, aluminium, fibreglass; h. 78.2cm (30$\frac{3}{4}$in) w. 70cm (27$\frac{1}{2}$in) d. 88.3cm (34$\frac{3}{4}$in). Frighetto Industrie srl, Italy

Below: Fabio Novembre. Combinable seating system, And. Polyurethane, wood, metal; h. 230cm (90$\frac{1}{2}$in) w. 175cm (68$\frac{7}{8}$in) l. 160cm (63in). Cappellini SpA, Italy

Fabio Novembre's 'And' took pride of place at the entrance of Cappellini's exhibition at Superstudio Piu during Milan, 2002.

At once retro and organic, the multifunctional tunnel was inspired by a backward interpretation of a string of DNA molecules.

The structure both divides and defines a space. Once seated within the labyrinth, ambient sounds are muted yet the gaps

between elements allows communication with the surroundings.

Jane Worthington. Sofa, DS 152. Leather, chrome; h. 71cm (28in) w. 193cm (76in) d. 156cm (61^{3}8in). de Sede AG, Germany

17 Futuretro

Fredrik Mattson. Seat, Innovation C. Lacquered steel, polyether foam, fabric;
h. 75cm (29¹2in) w. 70cm (27¹2in) d. 65cm (25⁵8in). Blå Station, Sweden

Chiara Cantono. Multi-purpose chair, E-Chair. Upholstered chair fitted with
Fujitsu Siemens Lifebook computer. Brunati Italia, Italy

Jean Marie Massaud. Chair, Lola. Injection-moulded plastic, aluminium; h. 80cm (31^{1}2in) w. 61cm (24in) d. 52cm (20^{1}2in). Liv'it srl, Italy

Opposite: Benjamin Hopf and Constantin Wortmann. Chair, Wop-chair. Fibreglass, steel; h. 75cm (29^{1}2in) w. 55cm (21^{5}8in) l. 55cm (21^{5}8in). Prototype; Büro für Form, Germany

Opposite: Philippe Starck. Studio chair, Hula Hoop. Polypropylene; h. 77.4–90cm (30^{1}2–35^{3}8in) w. 67.7cm (26^{5}8in) l. 67.2cm (26^{3}8in). Vitra AG, Switzerland

Above: Christopher Streng. Stool, NaLi. Reinforced plastic, steel, aluminium; h. 76cm (30in) w. 35.5cm (14in). Limited batch production; Christopher Streng Inc, USA

Shosaku Kawashima. LCD projector, Canon LV-X1/S1. ABS, acrylic, rubber, glass, steel and others; h. 7.6cm (3in) w. 26cm (10^{1}4in) d. 23cm (9in). Design Center, Canon Inc, Japan

Opposite: Achim Heine. Digital camera, Leica Digilux 1. Magnesium; h. 8.3cm (3^{1}4in) w. 12.7cm (5in) d. 6.7cm (2^{5}8in). Leica Camera AG, Germany

Opposite: Perry King and Santiago Miranda. Portable dehumidifier, Seccoreale Elettronico. Plastic; h. 45cm (17³⁄₄in)
w. 33cm (13in) d. 27cm (10⁵⁄₈in). King-Miranda Associati, Italy

Above: Tom Karen. Transistor radio, Bush TR130 (relaunched). Bush, UK

The TR130 was the UK's best-selling radio in the 1960s. Bush relaunched this model in time for the 2002 World Cup.
In 1966, when England was victorious, thousands of people heard the immortal lines 'They think it's all over. It is now.'
on their transistor radio.

Opposite: Mario Cananzi. Armchair, Twin. Metal tube, rattan core. Vittorio Bonacina, Italy

Below: Fabiano Trabucchi. Day bed, Tambao. Lacquered epoxy steel frame, chrome finishing, natural cane with wax treatment, polyurethane foam; h. 72cm (28^{3}8in) w. 120cm (47^{1}4in) l. 180cm (70^{7}8in). Bonacina Pierantonio, Italy

Jean Marie Massaud. Armchair, Ice Babe. Rotoplastic; h. 45cm (18in) w. 92cm (36¼in) d. 81 (31⅞in). Liv'it srl, Italy

Richard J. Anuszkiewicz. Art Borders wall covering, 1752. Marburg Tapetenfabrik, J.R. Schaefer GmbH & Co. KG, Germany

Opposite: Patrick Jouin. Chair, FOL.D. Polypropylene, chrome tubular steel. xO, France

Below: Ronan and Erwan Bouroullec. Folding armchairs. Nickel-coloured epoxy satin lacquer coated steel, PVC, polystyrene, EPDM; h. 71.7cm (28^{1}4in) w. 78.1cm (30^{3}4in) d. 69.2cm (27^{1}4in), high back armchair: h. 99.7cm (39^{1}4in). Ligne Roset, France

Thomas Sandell. Chair, IKEA PS Vägö. Polypropylene; h. 71cm (28in) w. 74cm (29⅛in) d. 92cm (36¼in). IKEA, Sweden

Guglielmo Berchicci. Armchair, Jetsons. Fibreglass, steel, aluminium, polyurethane, protective fabric; h. 110cm (43³⁄₈in)
l. 110cm (43³⁄₈in). Giovannetti srl, Italy

Eero Aarnio. Rocking chair. Fiberglass, steel tube; h. 105cm (43³⁄₈in) w. 67cm (26³⁄₈in) d. 127cm (50in). Artekno-EPS OY, Finland

Not since Gillo Dorfles, who followed on from Herman Broch and Clement Greenberg, has kitsch been really critiqued in 'high design'. Designers, the purveyors of 'good taste', spent a century trying to create a condition of clean, beautiful objects based on doctrines of proportions, scale, aesthetics, order, performance and material. Trying to design within these accepted 'tasteful' idioms and work to the modernist axiom of 'design' meant avoiding the figurative, the literal, and not creating or using languages that were considered cheap, low-brow or poor replicas of the 'authentic'. Kitsch is the result of a work of art that is transferred from its original meaning and used for a different purpose from that for which it was created — a remote circumstance, event or thing that is adopted and 'plagiarized'. A sense of ennui with Modernism, so-called 'good taste', 'elite design' or 'timeless' has fermented an underground movement of revival that is instilling objects with other meanings, one-liners, humour or intentional 'bad taste'. I believe the digital age has created this new autonomy — 'no class', 'no-brow', 'no original' — as symbolic of the fact that kitsch can be redefined and coexist in our object culture. Since we have no original, can we distinguish or even recognize good and bad taste? There maybe no measures, or in the words of Nietzsche, it may now be 'beyond good and evil'.

Nukitsch

'Beauty is not a quality inherent in things: it only exists in the mind of the beholder.'
David Hume

Previous page: Karim Rashid. Ego & Id. Clear and frosted glass; h. 40cm (15³⁄4in) diam. 30cm (11⁵⁄8in). Vitrocristal, Portugal

Above: Antonio Cos. Heart of Bottles. Glass; h. 18cm (7in) w. 25cm (9⁷⁄8in) d. 7.8cm (3in). One-off

Opposite: Stefano Giovannoni and Elisa Gargan. Cotton pad dispenser, Pill Up. PMMA; h. 23.5cm (9¹⁄4in). Alessi SpA, Italy

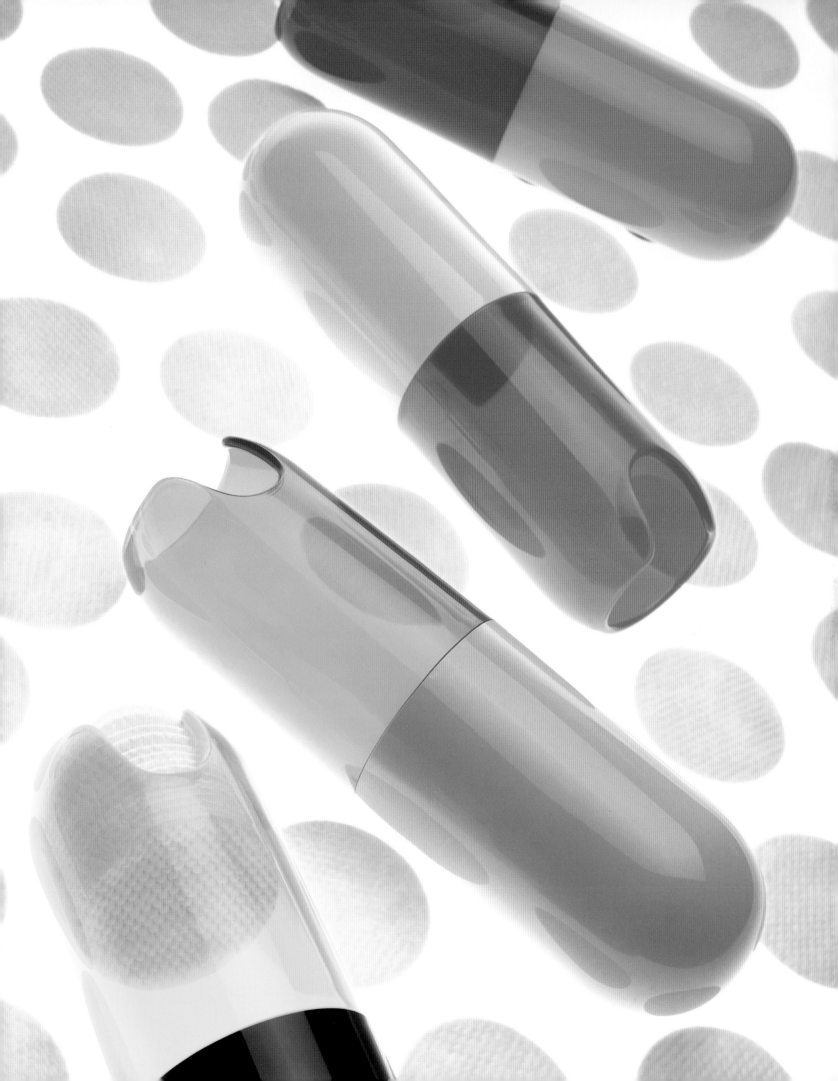

Marcel Wanders. Lamp, B.L.O. Plastics; h. 20cm (7⁷8in) w. 10cm (4in) d. 10cm (4in). Flos, Italy

Opposite: Gabrielle Lewin and Hlynur Vagn Atlason. Wastebasket, Flame. Acrylic; h. 40.6cm (16in) d. 26.7cm (10¹2in). Prototype

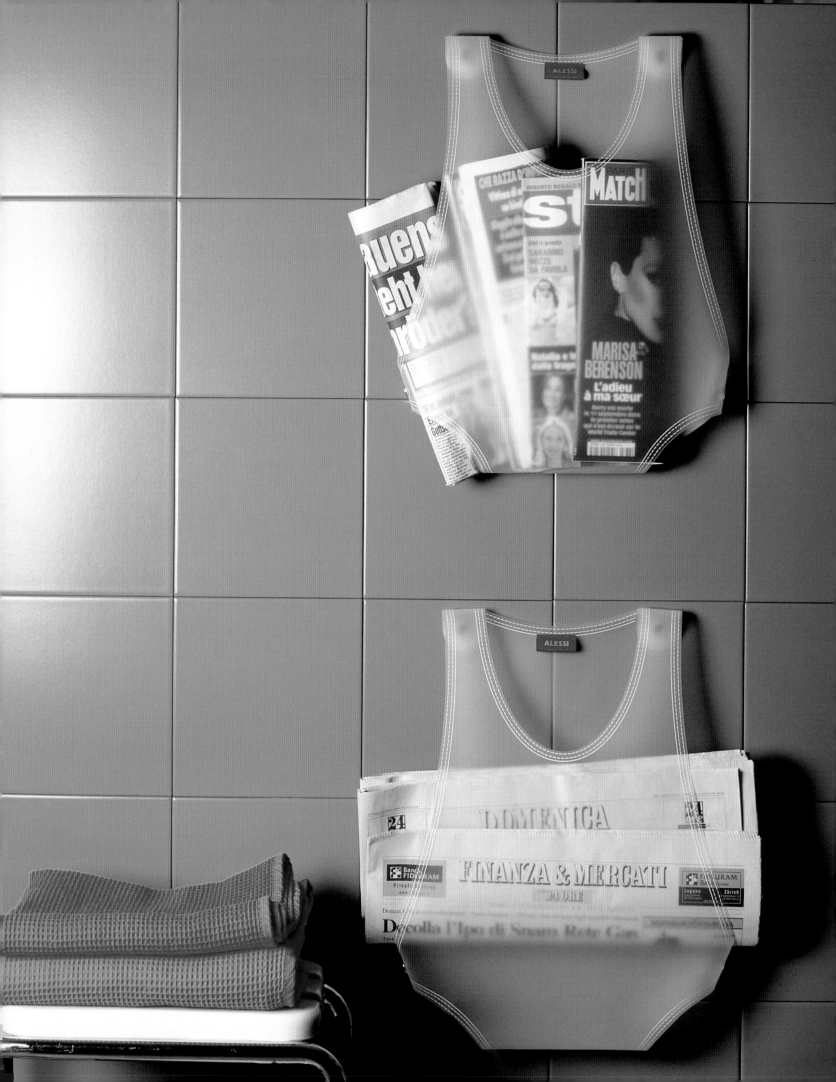

Opposite: WAAC's. Wall magazine holder, Marcel. W. 44.5cm (17$\frac{1}{2}$in) l. 36cm (14in). Alessi SpA, Italy

Anna Citelli. Lamp, Taglia Elle. PVC; h. 50cm (19$\frac{5}{8}$in) w. 40cm (15$\frac{3}{4}$in)

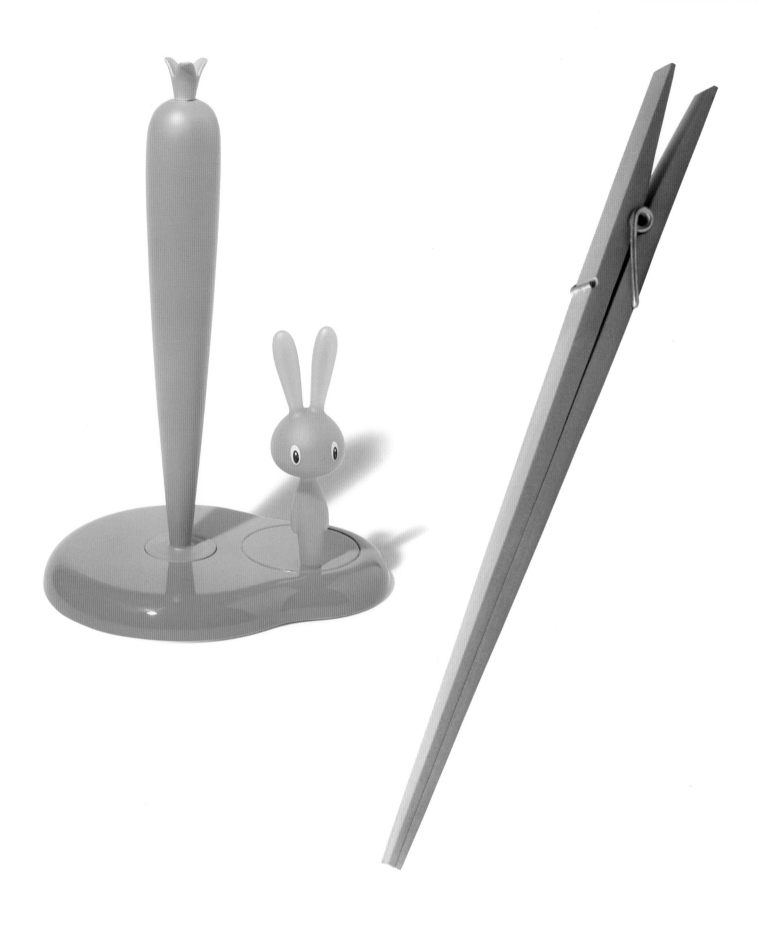

Stefano Giovannoni. Kitchen roll holder, Bunny & Carrot. Thermoplastic resin; h. 23.4cm (9¹⁄4in) w. 20.2cm (7⁷⁄8in) l. 16cm (6¹⁄4in). Alessi SpA, Italy

Samuel Accoceberry and Antonio Cos. Chopsticks, Joy-Stick. Beechwood; h. 20cm (7⁷⁄8in) w. 1.1cm (³⁄8in) d. 1.1cm (³⁄8in). Prototype

Alessandro Mendini. Large vase, Vaso Viso. Ceramic; h. 110cm (43³⁄8in). Alessi SpA, Italy

Yoshitomo Nara. Ashtray, Too Young To Die. Porcelain; h. 3.2cm (1¹⁄4in) d. 25.4cm (10in). Bozart, USA

Meike van Schijndel. Urinal, Kisses. Plaster and paint; h. 50cm (19⅝in) w. 35cm (13¾in) d. 35cm (13¾in). Prototype

Karim's sense of fun and mischief would not let him exclude this perfect example of kitsch. It's even better that this 'sexy' mouth has been designed by a woman as a reaction to the usual white sanitary wares that don't appear to have changed for centuries. Stefano Giovanonni's re-working of the urinal (see page 101) being the exception that proves the rule.

Stefano Giovannoni. Hot-water bottle, Hot Bott. Natural rubber; w. 21.5cm (8¹⁄2in) l. 30.5cm (12in). Alessi SpA, Italy

Eero Aarnio. Seat/sculpture, Chick. Foam, polyurethane; h. 80cm (31^{1}2in) w. 60cm (23^{5}8in) l. 100cm (39^{3}8in). Prototype

Eero Aarnio. Seat, Pony. Polyurethane foam, steel tube; h. 70cm (27^{1}2in) w. 60cm (23^{5}8in) l. 100cm (39^{3}8in). Adelta, Germany

Eero Aarnio. Marionette, Spaceman. Birch, acrylic; h. 50cm (19^{5}8in) w. 25cm (9^{7}8in) d.15cm (5^{7}8in). Prototype

When I asked Aarnio, as a designer, what role he would like to play in today's society, he replied, 'I play no "role" at all. When designing these items I have been very honest to myself, I'm transparent, my designs are like my signature, they can be executed only one way, my way. I have no fine explanations for my works; I just design items I like. To answer briefly, I love to create.' Probably most famous for his hanging 'Bubble' chair of 1968, Aarnio was a pioneer in the use of plastics, which invaded the market during the mid-1960s, reigned supreme until the mid-1970s, and are now enjoying renewed popularity. Plastic set designers free to make any shape or use any colour. The resulting creations were both fun and functional. Aarnio was born in Helsinki in 1932 and studied at the Institute of Industrial Arts. He became a household name following his collaboration with the Asko manufacturing company, who shared his vision that design means 'constant renewal, realignment and growth'. The organic form of the 'Globe' chair liberated Scandinavian design from its reputation for serene minimalist elegance. Here was a design that was fun and futuristic (the chair came complete with built-in telephone or stereo speakers), and summed up the swinging 1960s. Today Aarnio works mainly for Adelta, who manufacture re-editions of the famous pieces as well as his new designs. The 'Pony' chair is a re-issue of a 1970 classic, and is constructed in polyurethane foam. Aarnio comments, 'My cold-foamed designs get people to laugh and remember the best time of their childhood. They grow younger as they rediscover how funny and exciting it is to play. Inside of every person there is still a little boy or girl who is curious and creative. I hope these designs will survive in the same way that blowing a soap bubble interests every new generation.'

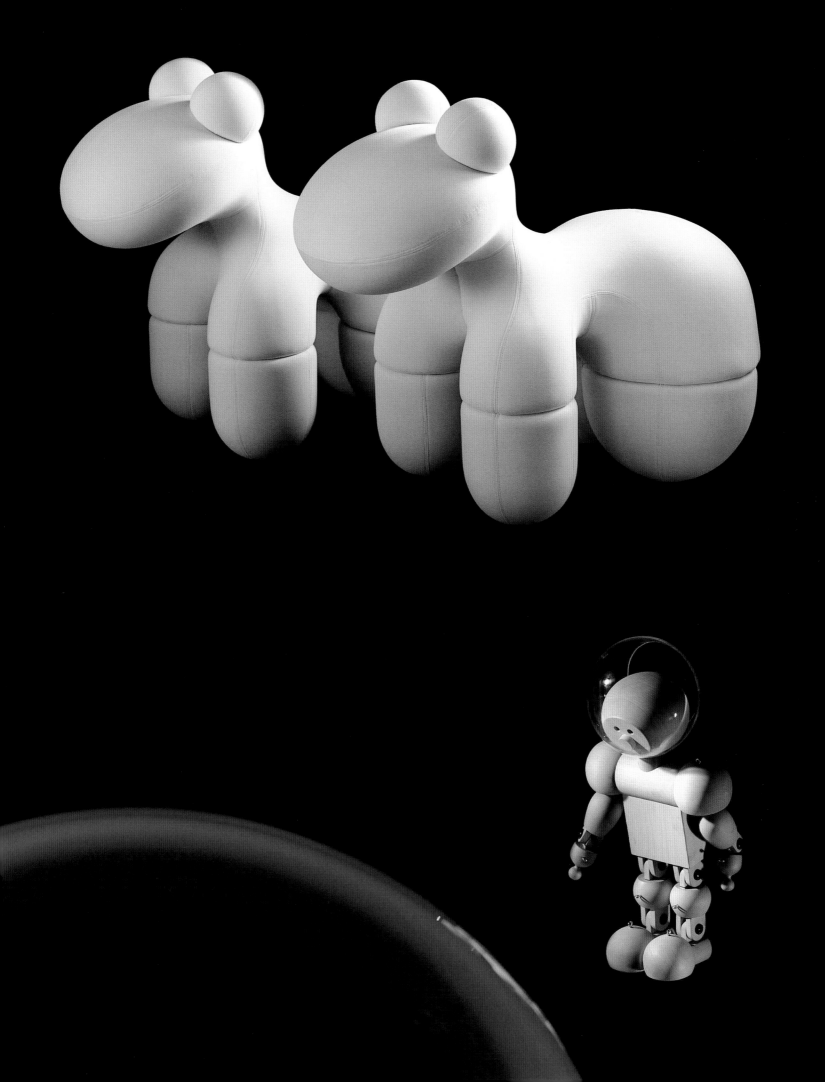

Stefano Giovannoni. Electric shoe polisher, Benny Brush. ABS, metal; h. 50cm (19³4in) w. 30cm (11³4in) l. 60cm (23⁵8in). Elmar Flototto GmbH, Germany

Opposite above: Panasonic Design Company. Headphone stereo, RQ-CWO5. ABS, steel; h. 9.4cm (3³4in) w. 11.6cm (4¹2in). Matsushita Electric Industrial Co. Ltd, Japan

Opposite below: Panasonic Design Company. Soft iron, NQ-SP10. ABS, steel; h. 13cm (5in) w. 9cm (3¹2in) l. 21cm (8¹4in). Matsushita Electric Industrial Co. Ltd, Japan

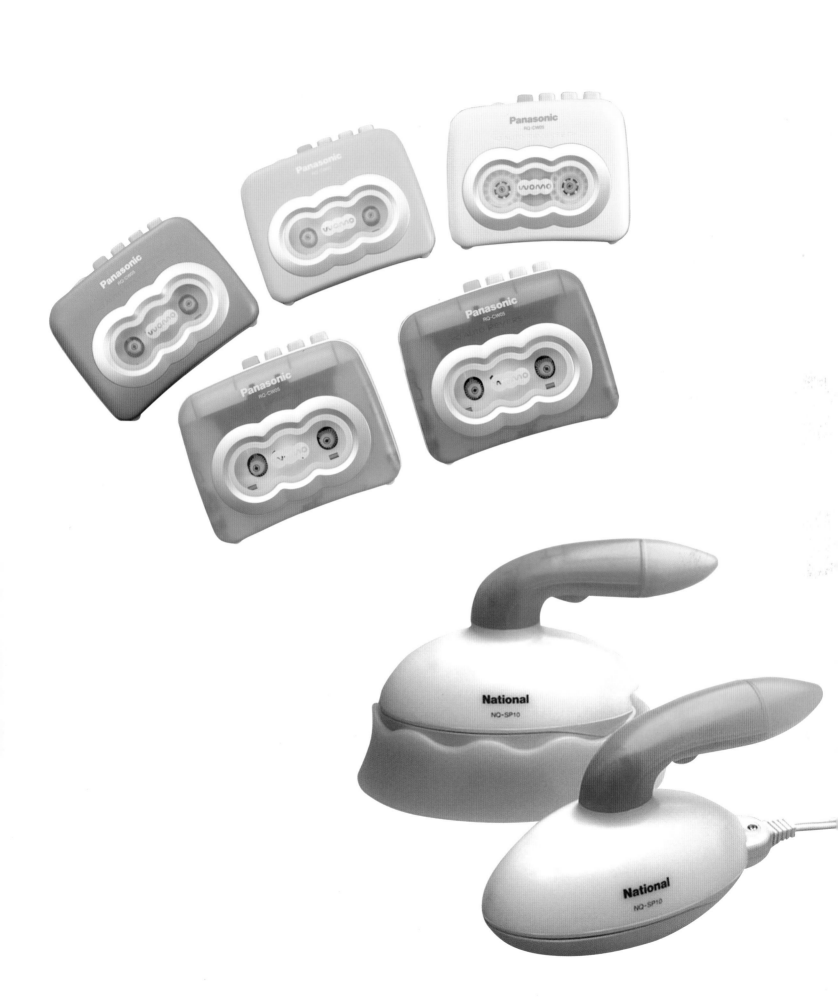

Constantin Boym. Chair, Taxicab. Metal frame, wooden balls; h. 91.4cm (36in) w. 40.6cm (16in) d. 45.7cm (18in). Boym Partners, USA

Karim chose much of what is in this book blind – he did not want to know who the designer was or what we were looking at. When shown this chair he immediately said, 'New York taxi'. For anybody living in that city a seat of massage beads could evoke nothing else. Boym, as a reaction to minimalism, has developed a collection that is 'more intense, more inclusive, and a bit more fun'. His exhibition concerned itself with decoration. Instead of being placed on a chair, the beads become the seat itself. Each was picked manually by one of Boym's colleagues, who insisted on being blindfolded so as to create a totally random pattern.

Daniela Polubedovová and Stanislav Fiala. Seat, Kurnik Shopa. Tennis balls, perspex, steel. One-off

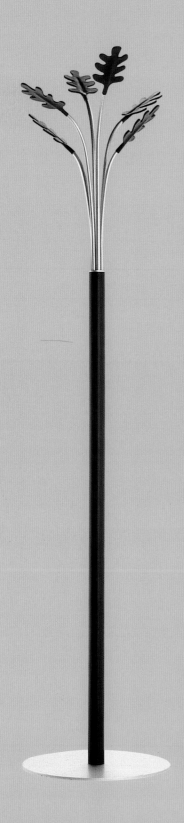

Michele de Lucchi and Philippe Nigro. Coat stand, Tri. Metal, leather; h. 172 cm (67³⁄4in) w. 40cm (15³⁄4in) d. 40cm (15³⁄4in).
Poltrona Frau srl, Italy

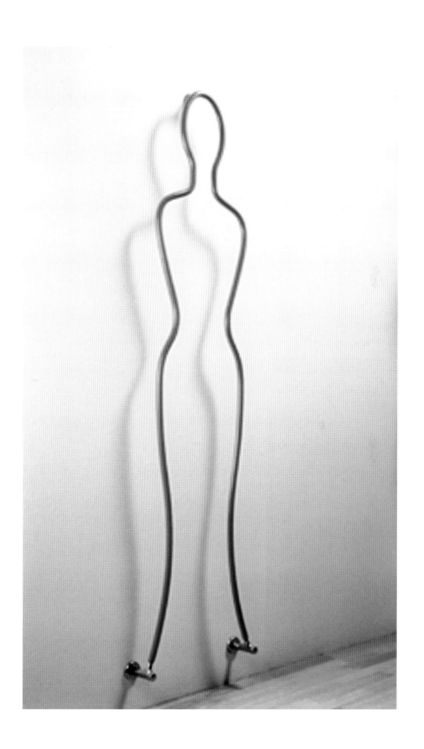

Paolo Pedrizzetti. Bathwrap warmer, Miss Hot. Steel; h. 176cm (69¹3in) l. 42cm (16¹2in). Tubes Radiatori, Italy

Above: Rainer Spehl. Stool, Qoffee. Polyethylene; h. 47cm (18^{1}2in) d. 35cm (13^{3}4in). Wydale Plastics, UK

Opposite: Constantin and Laurene Leon Boym. Salvation ceramics. Styling and realisation: Rebecca Wijsbeek. Secondhand porcelain; h. 25cm (9^{7}8in) diam. 15cm (5^{7}8in) or h. 35cm (13^{3}4in) diam. 25cm (9^{7}8in) or h. 50cm (19^{3}4in) diam. 30cm (11^{3}4in). moooi©, The Netherlands

Above: Gaetano Pesce. Armchair, Nobody's. Polyurethane resin. Zerodisegno, Quattrocchio, Italy

From the manner in which Rashid has grouped many of the designs in the Yearbook, it is revealing how many designers are working in a similar vein. Nevertheless, there comes across a clear reaction to the homogeneity and uniformity of minimalism, giving the selection a heart and emotion. Zerodisegno, in collaboration with Gaetano Pesce, has developed a line called 'Nobody's Perfect' as an expression of the diversities of people and regions of the world. Each item is unique, cast in a multicoloured resin that sometimes completely fills the mould and sometimes not, creating anthropomorphic shapes that are not fully formed until they are released and harden. Each piece includes a date that has been hand-cast inside its own skin, and every example, in its own 'imperfection', represents something that is unique, different and extraordinary: each one contains an act of creativity.

Opposite: Juan Benavente Valero. Lamp, Light-Pot. Polypropylene. Prototype; Juanico Design, Spain

Above: Anna Citelli. Armchair, Poltiglia. PVC covering, plastic bottles; h. 100cm (39³8in) w. 100cm (39³8in) d. 100cm (39³8in)

Opposite above: Jerszy Seymour. Armchair, Muff Daddy. Denim; h. 65cm (25¹2in) w. 105cm (41³8in) l. 110cm (43¹4in). Covo, Italy

Jerszy Seymour's 'Muff Daddy' is a twenty-first century reinterpretation of the casual living concept of the 1960s, of not taking the world too seriously. Design should adapt to how we conduct our lives; this armchair exemplifies 'cool' lounging. Seymour predicts that, 'The future of design is going to be fabulous, furry, furious and fun. It will be a tool of love, a superhero ready to do battle for good and evil. It will ask why it exists and what its purpose in life is. It will stick its middle finger up, run through the woods naked, and save the world.'

Opposite below: Monkey Boys. Chair, Soft. Metal, polyurethane foam, fabric; h. 80cm (31¹2in) w. 60cm (23⁵8in) d. 60cm (23⁵8in). Prototype

WE DO NOT NEED MANY THINGS. IN FACT, ALL THAT IS NECESSARY ARE SIMPLE OBJECTS THAT HAVE MEANINGS THAT ARE SACRED TO OUR LIVES. THEY ARE FEW AND PERSONAL. TRAVELLING THROUGH THE WORLD IN OUR POST-INDUSTRIAL AGE, I CANNOT HELP BUT SEE A GLOBAL SHRINKING WHERE CULTURES, PRODUCTS AND SIGNS ARE BECOMING SIMILAR, LANGUAGES ARE CONVERGING, AND UNIVERSALITY IS A PRESENT PHENOMENON. THE CONVENIENCE AND RELIANCE OF WHAT ALREADY EXISTS ALL TOO OFTEN BECOMES NUMBING, AS BLIND REPETITION CREATES DULLNESS, AND ENNUI ENSUES WHERE BANALITY EQUALS ROUTINE. WE LIKE ROUTINE BECAUSE IT IS EASY, COMFORTABLE AND ACCOMMODATING. IT TAKES GREAT EFFORT TO BREAK OUT OF IT AND HAVE NEW EXPERIENCES. GENERALLY WHEN ONE EXPERIENCES CHANGE IT BECOMES AN 'AWARENESS' OF TIME, AN UNVEILING OF AN ACTUALITY. THIS PHENOMENOLOGICAL MOMENT OF SURPRISE ELEVATES OUR SENSE OF LIFE AND BEING. WHY DO I SHOP WHEN I TRAVEL? IT SEEMS LIKE THE ONLY REASON IS TO SEARCH FOR MORE EXCLUSIVE LOCAL GOODS THAT ARE NOT AVAILABLE ELSEWHERE. SO IT IS THE PHYSICAL EXPERIENCE, THE TOUCH, THE SMELL, THE TASTE, THAT WILL CONTINUE TO BRING US TOGETHER WITH DIVERSITY. THE OBJECTS ACT AS CATALYSTS. OURS IS AN AGE OF TECHNOLOGICAL HYPERTROPHY THAT, DESPITE THE EXPLOSION OF OPTIONS AND FUNCTIONALITY IT HAS CREATED, THREATENS INDIVIDUALITY. IN THIS BREACH OF NUMBING UNIFORMITY, DESIGNERS CAN COMPOSE THE IDIOMATIC AND VARIABLE, THINGS THAT CHANGE OVER TIME, REVEAL NEW PHENOMENA AND INCREASE OUR AWARENESS OF THE EXPERIENCE, MAKING OUR PHYSICAL WORLD NOT ONLY MORE POTENT AND HUMANE, BUT MORE TRANSCENDENT AS WELL. FOR PLAYING UPON THE DIVERSITY AND RICHNESS OF 'CHANGE OVER TIME' IS A HOMMAGE TO THE INDIVIDUAL AND THE SENSES – A SURPRISING OF THE SOUL.

PHENOMENA

'THE RESULTING STRONGLY FELT CHANGES IN OUR PERCEPTION, IN OUR RELATIONSHIP WITH THE WORLD, AND IN OUR EVERYDAY BEHAVIOUR ATTEST TO A BIRTH OF NEW AESTHETICS. IT IS AN AETHETICS WHOSE DESIGNATED OBJECT LIES BEYOND THE VISIBLE, BEYOND THE TANGIBLE, IN THE ZONES OF INFRAPERCEPTION, ALONGSIDE OUR MODERN AWARENESS.' FRED FOREST

Previous page: Karim Rashid. Chess set. Thermoplastic elastomer (TPE), acrylic; board: w. 56cm (22in) l. 37cm (14^{1}2in), pawn: h. 6cm (2^{3}8in), knight: h. 9.5cm (3^{3}4in), queen: h. 10cm (3^{7}8in), king: h. 10.5cm (4^{1}8in). Bozart, USA

Above: Agnoletto-Rusconi. Standard lamp, Asa. Steel, epoxy powders, steel plate, incandescent bulb; h. 138/168cm (54^{3}8/66^{1}8in). Palluccoitalia SpA, Italy

Ferdi Giardini & Associati. Table lamp, Nerolia. Transparent crystal, amber glass, dimmer; h. 30cm (11⁷⁄₈in) diam. 12cm (4³⁄₄in). Oluce srl, Italy

Dimmed lighting and relaxing aromatherapy combine in a modern reworking of an ancient ritualistic ceremony. As the lamp produces warmth the scents are released into the atmosphere.

Opposite: Markus Benesch. Top: Bench, Cube-tube. Acrylic glass, lenticular/photopaper image, T5 dimmable neon light, wood, silversurfer laminate; l. 140cm (55$\frac{1}{8}$in) diam. 30cm (11$\frac{7}{8}$in). Small table, Fonsino. Acrylic glass, lenticular/photopaper image, T5 dimmable neon light, wood, silversurfer laminate; h. 15cm (5$\frac{7}{8}$in) w. 60cm (23$\frac{5}{8}$in) l. 60cm (23$\frac{5}{8}$in). Bottom: Table, Straw. Acrylic glass, lenticular/photopaper image, T5 dimmable neon light, wood, silversurfer laminate; h. 15cm (5$\frac{7}{8}$in) w. 60cm (23$\frac{5}{8}$in) l. 180cm (70$\frac{7}{8}$in). Benesch Project Design, Germany

Markus Benesch's 'Cubes' act as a table, a stool, a lamp, or all three simultaneously. When switched off, they revert to silvery shining sculptures, an effect caused by the laminate he developed in 1997 specially for his designs.

Above: Elizabeth Paige Smith. Table, Blow. Acrylic, powder pigment; h. 50.8cm (20in) w. 45$\frac{3}{4}$cm (18in) d. 45$\frac{3}{4}$cm (18in). One-off

The translucence of the 'Blow' table is the result of static charges that build up between the acrylic box construction of the table and the loose powdered pigment resin it contains.

Opposite: Peter Christian. Shelves, Continental. Lacquered MDF, acrylic; h. 180cm (71in) w. 34cm (13^{1}2in) l. 154/81 cm (60^{5}8/31^{7}8in). Christian Stuart Partnership, UK

Above: Peter Christian. Pendant lights, Shimmer. Acrylic, aluminium; diam.15cm (5^{7}8in). Aktiva, UK

Peter Christian has been experimenting with lighting effects since he left the Royal College of Art in 1984. A thin layer of optical film trapped behind an acrylic diffuser reflects a myriad of colours, which appear to alter depending on the angle of observance. When the light is extinguished they are still apparent but are muted, the overall impression becoming more silvery.

Jensen Holbak. Self-watering planter, Eva Solo. Eva Denmark, Denmark

Jason Miller. Vase, Bloom, Mouth-blown glass, photochromic pigmented ink; h. 27.9cm (11in) w. 10.2cm (4in). Limited batch production; Jason Miller Studio, USA

'Bloom' is a series of vases that literally flower when exposed to sunlight. Printed with photochromic ink, the graphics are white but change to a colour in the presence of sunlight.

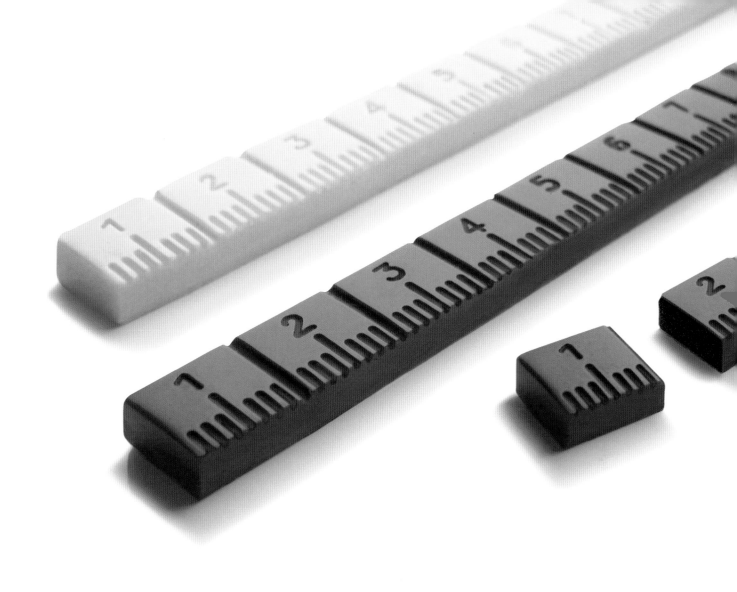

Opposite: Renaud Thiry. Chocolate. H. 1.2cm (^{1}2in) w. 5.4cm (2^{1}8in) l. 5.4cm (2^{1}8in). Patisserie Pulicani, France

Above: Pablo Ulian. Chocolate, Greediness Metre. White and fondant chocolate; h. 1cm (^{3}8in) w. 2.5cm (1in) l. 40cm (15^{3}4in). Prototype

Paolo Ulian describes his metre-long novelty chocolate bar as a combination of two typologies that measure greed.

Ilaria Marelli. Tables, Apple. Plexiglas, polyethylene; small table: h. 41cm (16^{1}8in) w. 48cm (18^{7}8in) d. 48cm (18^{7}8in), low table: h. 27cm (10^{5}8in) w. 63cm (24^{3}4in) l. 115cm (45^{1}4in). Cappellini SpA, Italy

Scott Henderson. Personal carafe, Wovo. Projection-moulded propylene, formed stainless steel, projection-moulded SAN; h. 18.5cm (7^{1}4in) diam. 8.5cm (3^{3}8in). Wovo, USA

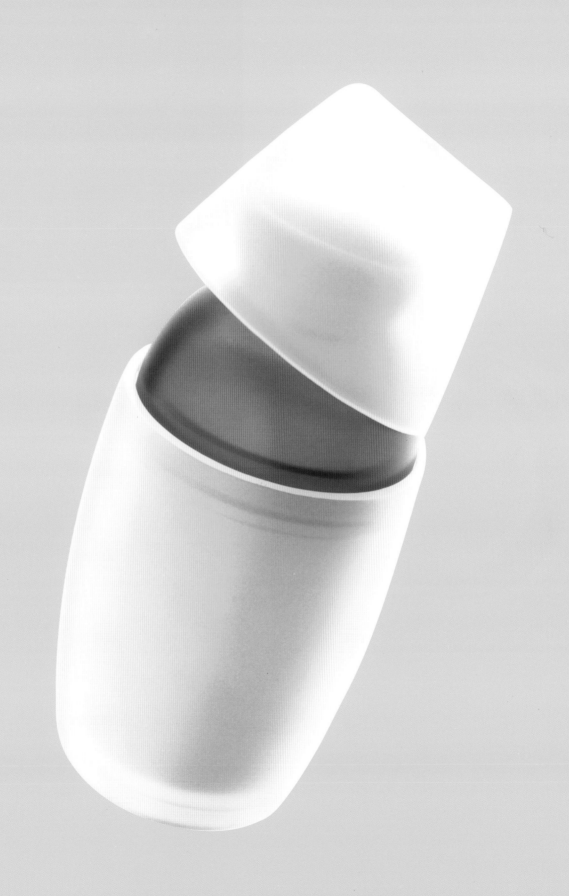

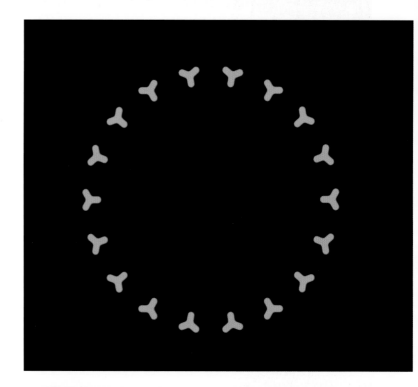

Top left and right: Andrea Ruggiero. Wall-mounted mirror, Glo. Luminova, mirror, MDF; diam. 64cm (25^{1}8in). Limited batch production; Bart Design, Italy

Above left: Shin and Tomoko Azumi. Lamp, NW11 (Ring Light). Span aluminium, fluorescent lamp; h. 7cm (2^{3}4in) diam. 62cm (24^{1}2in). Mathmos, UK

Above right: Andreas Winkler. Door handle, Top. Brushed stainless steel and porcelain; l. 13cm (5^{1}8in) d. 7cm (2^{3}4in) diam. 1.8cm (3$^{}$4in). Phos Design, Germany

Opposite: Denis Santachiara. Lamp, Mister Tesla. Tesla coil, fluorescent tubes, glass, ceramic; h. 90cm (35^{3}8in) w. 45cm (17^{3}4in) diam. 24cm (9^{3}8in). Memphis srl and Post Design, Italy

To design for the sake of form or to suit the demands of the market are alien concepts for Denis Santachiara. He believes that all objects should be based on an ideology, or have a connotative language, without which they will be flat and superficial. Designs may be well executed, but they will never be expressive, or rise to the poetic. His work has consistently shown his obsession with invention and a desire for new experiences, and is always ahead of its time. In an interview for 'Intramuros' magazine he writes that seventeen years ago he held a one-man show entitled 'Neo-Merchandise Design of Invention and Artificial Ecstasy' at the Milan Trienalle. In an accompanying catalogue, Santachiara composed a glossary for the text he had written on the products shown, all of which had been designed with the latest ironic and 'performance' technologies. The glossary contained words such as 'interactive', 'virtual', 'high-performance', 'gadget', 'functionnoid', 'surprising', 'magical' and 'immaterial', all very forward-looking and of much relevance today, as demonstrated in the selection illustrated in this book. To Santachiara, design is dying because too many are working without new visions, and because manufacturers are playing it safe. Inspiration should not just come from the design world but from other creative disciplines and forms of expression. His 2002 Milan show 'Sorry for the Plug' was dedicated to the nineteenth-century Croatian scientist Nikola Tesla, a pioneer in radiophonic experiments and the inventor of the alternating current, the first cathode tube and the prototype of the electronic microscope. Using luminescent tubes that produce X-rays, he photographed his own hands under wireless fluorescent lights. It is not the inventions themselves, however, that inspire Santachiara, but the visionary nature of the concepts and the amazement Tesla manifested in front of his own work. He described the effect of the microscope as being 'magnificent, a wonderful vision, a tremendous display, glorious, so marvellous that somebody might be scared to talk about it'. It is this enthusiasm and the allusion to the magical that captivates Santachiara and which he wanted to recapture at the Post Design Gallery in 2001. The 'Mister Tesla' lamp installation consists of a small ceramic column upon which sits a glass jar containing neon lights, lit without the use of wiring. The tubes stay on even when the jar is being moved or handled. Added to this, a wall-fitted glass shelf switches on any neon light placed in its vicinity.

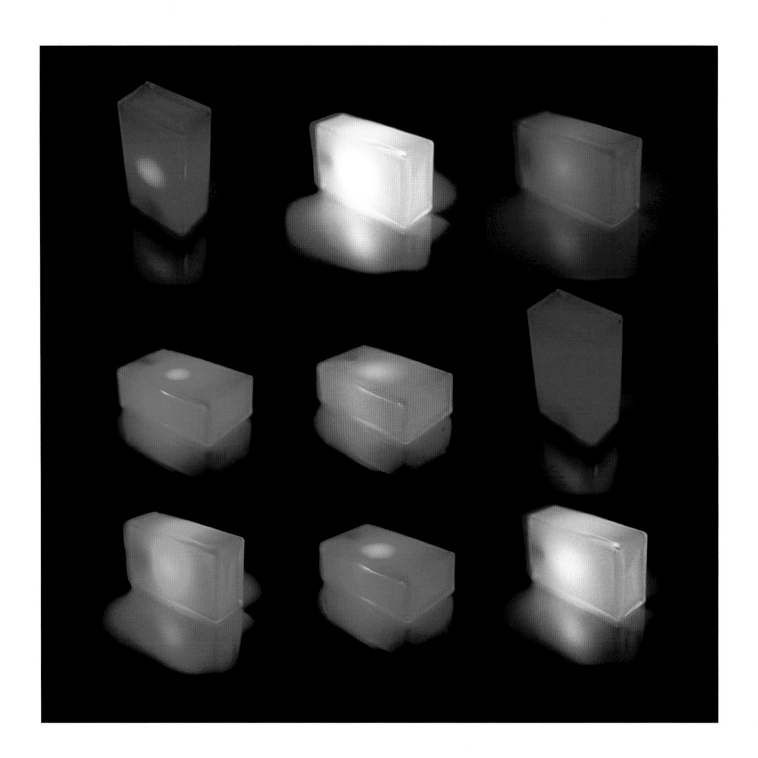

Mathmos in-house design team. Portable light, Tumbler. Glass, LEDs, recharger; h. 10cm (4in) w. 6cm (3¹4in) l. 18cm (7¹2in). Mathmos, UK

Following on from last year's innovative and successful 'Bubble' light, Mathmos have developed 'Tumbler'. Charged like a mobile phone, it can be used inside and out without the encumbrance of lead or socket. The first variation can be set either to pulse or to be static in red, green or blue lights, while 'Tumbler Changing Colour' incorporates PIC chips and sensors to allow the lamp to glow in different hues depending on which of its fascias it is placed on. A variant phases through seemingly infinite colour combinations, the speed altering from fast to barely perceptible.

Terri Pecora. Sink, Déjà-vu. Glass, wood, brushed steel; l. 105/160 cm (41$\frac{3}{8}$/63in) d. 57cm (22$\frac{3}{8}$in). Cristalquattro srl, Italy

Above: Mark Mann and Markus Huppmann. Necklace, LightBlueCollar. Glass, LEDs, wire, PP; h. 17cm (6³4in) w. 31cm (12in) d. 1.8cm (³4in). Prototype

LEDs are used to light antique pharmacist bottles in a poetic blend of old and new.

Opposite: Adrian Peach. Sofa, Playstation. Steel tube, upholstery foam, fabric; h. 55cm (21⁵8in) w. 160cm (63in) l. 120cm (47¹4in). Felicerossi srl, Italy

Overleaf left: Olivier Sidet (Radi Designers). Mirror, Ghost. Mirror, angled film; h. 80cm (31^{1}2in) l. 130cm (51^{1}8in). Glace Control, France

Using angled film within the mirror, 'Ghost' allows the observer to see his surroundings but never himself.

Overleaf right: Tech International Corp. and Winsmart International Development Ltd. Photocube Levitator. ABS, electromagnet, magnet; h. 21.6cm (8^{1}2in) w. 29.2cm (11^{1}2in) , l. 22.5cm (8^{7}8in). Tech International Corp., USA

The Photocube Levitator uses magnets to make its four-sided photo display revolve in mid-air; it comes replete with 'disco-lights'.

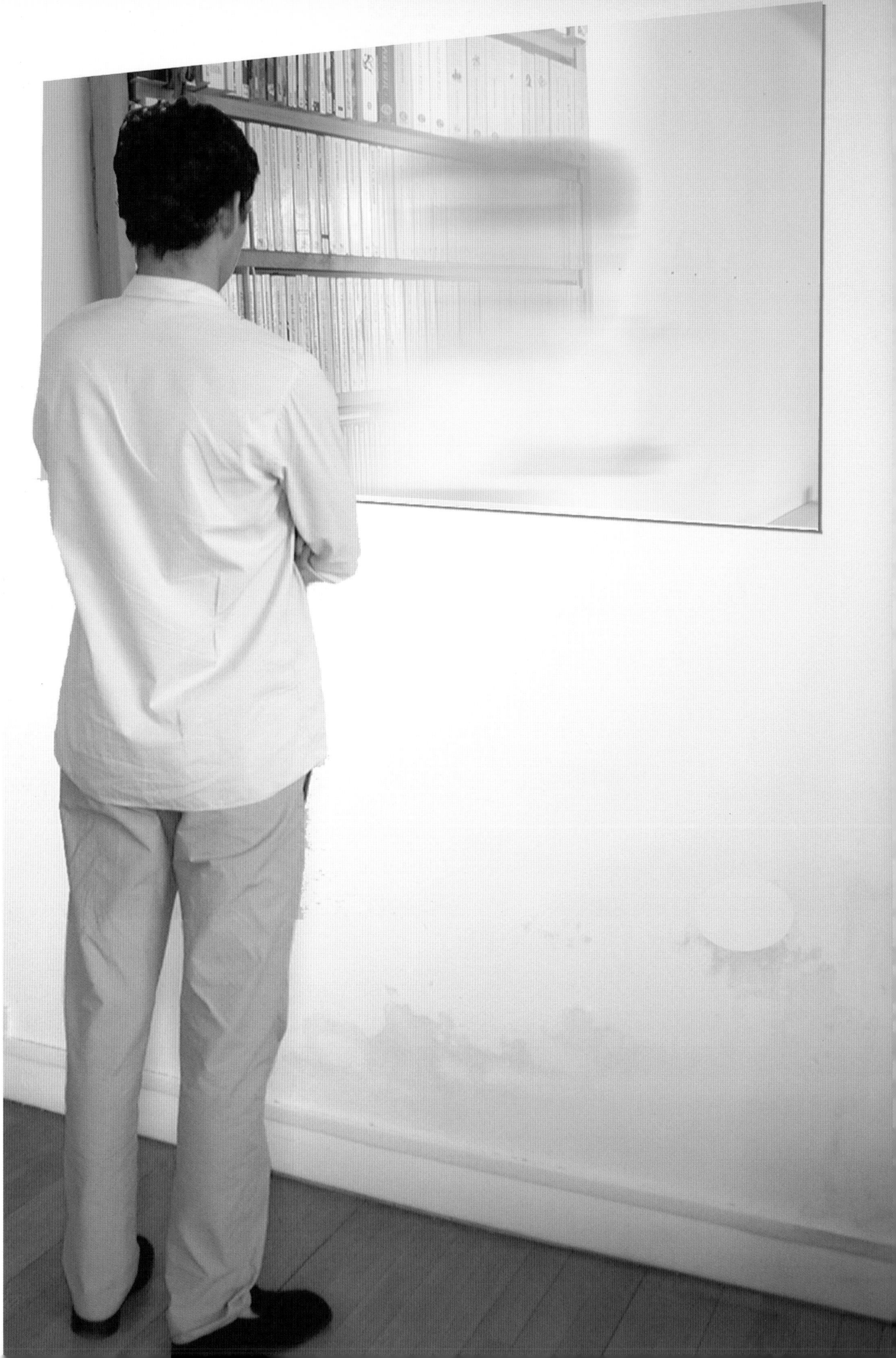

THE 'DIGITAL AGE', THE 'INFORMATION SOCIETY', THE 'GLOBAL VILLAGE' AND THE 'LEISURE CULTURE' ARE ALL PHRASES THAT DESCRIBE A CHANGING PHYSICAL WORLD WHERE 'SOFT', ORGANIC MUTABILITY CHARACTERIZES OUR LANDSCAPE. THE DESIGN 'RAPPINGHOOD' IS WHERE A PLETHORA OF SHAPES, MATERIALS, MECHANICS, ELECTRONICS, DIGITALIA, INFORMATION, MYTHS, MARKETS AND TRIBES ALL FORM AN EVER-CHANGING, PARABOLIC COSMOS THAT IS OUR BUILT ENVIRONMENT. IT COMBINES WITH THE HUMAN BODY, SOFTNESS AND TACTILITY TO COALESCE INTO EMERGING LANGUAGES, ALL THESE COMPONENTS GIVING FORM TO OUR NEW FORMAL LANDSCAPE THAT IS SOFT. IT IS AN AXIOM OF OUR NEW PHYSICAL CONDITION, A DESIRE TO CONQUER ALL MATHEMATICAL PROBABILITIES WITH FORMS THAT COULD PREVIOUSLY NEVER BE DOCUMENTED IN ANY RATIONAL FORMULAE. SUCH OBJECTS ARE IN PART A RESULT OF NEW COMPUTER-AIDED TOOLS OF MORPHING. NURBS, SPLINES, METABALLS AND OTHER BIO-SHAPING COMMANDS ARE FOSTERING A MORE RELAXED ORGANIC CONDITION – SOFTER, GLOBULAR AND BLOBULAR. THIS SOFTENING, OR AS I CALL IT, 'CASUALIZATION' OF SHAPE, FORM, MATERIAL AND BEHAVIOUR, IS DEFINITELY A MOVEMENT – A DIGITAL LANGUAGE OF BLOBIFYING OUR WORLD, PHYSICALLY AND IMMATERIALLY. THE 'BLOBJECT' MOVEMENT IS HERE THANKS TO A HANDFUL OF DESIGNERS WHO HAVE TAKEN A GREAT INTEREST IN ORGANIC FORM AND THE TECHNOLOGIES THAT ARE ALLOWING US TO MORPH, UNDULATE, TWIST, TORQUE AND BLEND OUR LANDSCAPE.

ORGANIC

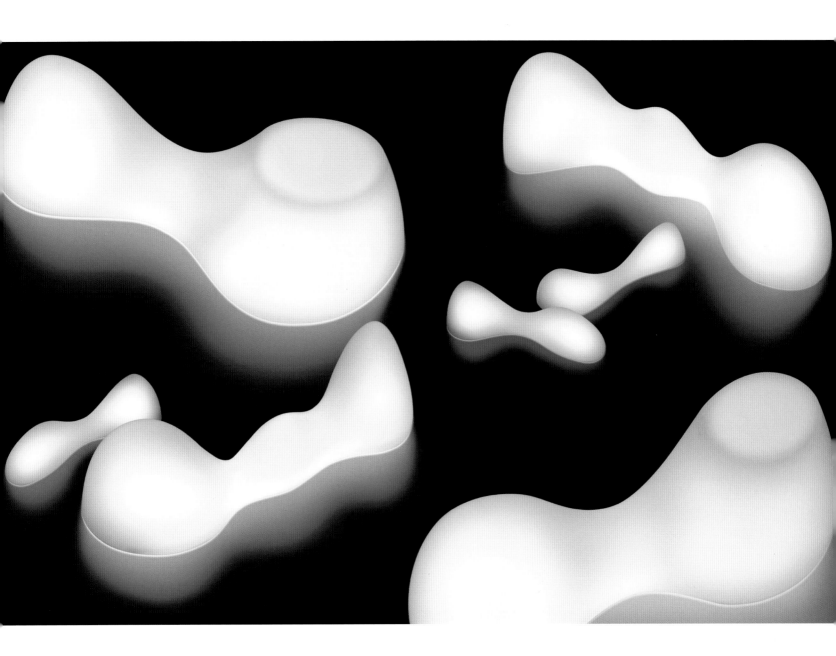

'The sublime, on the other hand, is to be found in a formless object, so far as in it or by occasion of it boundlessness is represented, and yet its totality is also present to thought.' Immanuel Kant

Previous page: Karim Rashid. Blob lamps. Acrylic; h. 15–51cm (5^{7}8–20in) w. 16–44cm (6^{3}8–17^{3}8in) l. 45–129cm (17^{3}4–50^{7}8in). Foscarini, Italy

Above: Kivi Sotamaa. Extraterrain. ABS, high density polythurethane, pearl white; h. 90cm (35^{3}8in) w. 280cm (110^{1}4in) d. 180cm (51^{1}8in). One-off; Ocean North, Finland

'Extraterrain' is urban landscape reduced to the scale of furniture. It has no pre-conceived usage but is defined by the inhabitor. The appearance of the complicated folded surfaces is without rigidity or stability, producing a continual reconfiguration of the space. It is thus described as a non-hierarchical arena for social, political and economic exchange.

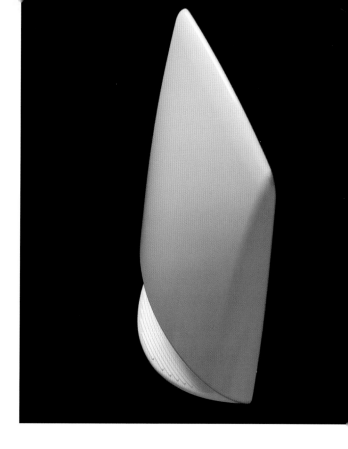

Top: PLH Design. Sanitary equipment, Drop. Urea plastic; h. 42cm (16^{1}2in) w. 30cm (11^{3}4in) d. 9cm (3^{1}2in). Panvision, Denmark

Above: Tokujin Yoshioka. Chair, Tokyo-pop. Polyethylene; h. 178.5cm (70^{1}2in) w. 74cm (29^{1}8in) d. 157cm (61^{7}8in). Driade SpA, Italy

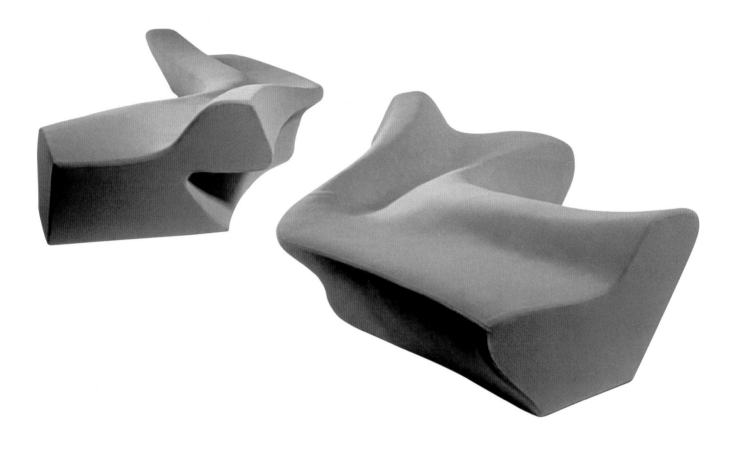

Zaha Hadid. Moraine Divan. Leather; h. 75cm (29^{1}2in) w. 250cm (98^{3}8in) l. 500cm (196^{7}8in). Sawaya & Moroni SpA, Italy

Jan Melis and Ben Oostrum. Seat, Pill. Polyethylene foam, rubberlike coating; h. 35cm (13^{3}4in) diam. 60cm (23^{5}8in). MNO, The Netherlands

Above: Kate Hume. Vases, Opal Rock and Pebble. Glass. Limited batch production; Kate Hume Glass, The Netherlands

Opposite: Jan Melis and Ben Oostrum. Seating object, Donut. Foam; h. 100 cm (39^{3}8in) w. 40cm (15^{3}4in). MNO, The Netherlands

Opposite above: Ross Lovegrove. Vases, Orgeramic study series one 2002. Glazed earthenware; h. 9cm (3^{1}2in) w. 26cm (10^{1}4in) l. 38cm (15in). Cor Unum, The Netherlands

Opposite below left: David Serero, Elena Fernandez and Philipp Mohr (Degre Zero Architecture). Metaring Bracelet. Glacier White DuPont Corian®; h. 5cm (2in) w. 10.2cm (4in) l. 10.2cm (4in). Sterling-Miller Designs, USA

Opposite below right: Allen Zadeh, Clay Burns, Mark Prommel and Ian Halpern. Burton Snowboard Tools and Tuning Line. Glass reinforced nylon, aluminium, stainless steel. Prototype; Burton Snowboards, USA

Above: Olivier Peyricot. Body Props. Polyurethane, elastic varnish. Edra SpA, Italy

Philippe Starck. Easy chair, S.T. Strange Thing. Titanium, fabric upholstery; h. 79cm (31in) w. 98cm (38^{1}2in) d. 76cm (29^{7}8in).
Cassina SpA, Italy

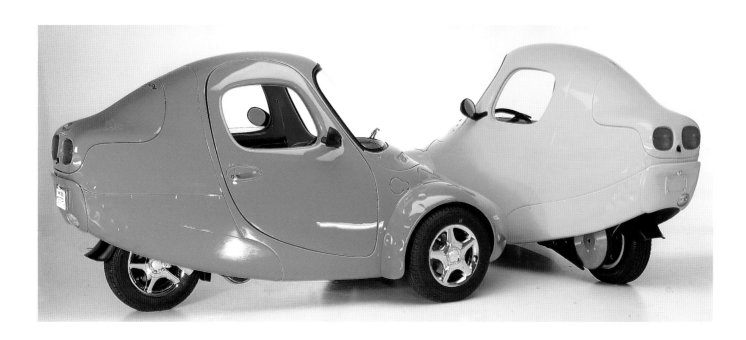

Mike Corbin. Electric car, Sparrow. Corbin Motors, USA

One of the ways humankind has progressed is through natural selection and adaptation. From this ability springs the very changes that alter how we live and that impact upon the future of the planet. One such transition will be in transportation. Environmental concerns and the need for smaller, more efficient vehicles have opened up the automobile market to one- and two-man units. The success of the Smart car attests to this, but Corbin Motors have gone one step further in producing the 'Sparrow' – not only is it small (121 x 244 cm, 4 x 8 ft) but it is electric and therefore less polluting. Mike Corbin first came up with the concept in the 1970s, inspired by his life-long love of motorbikes and his worries about the need for clear air and the price of petrol. In Corbin's design ethos, 'Easy Rider' (he engineered the custom look of the chopper motorbike) meets Captain Planet. He believes that clean fuel need not result in a lack of performance, and indeed he became the fastest man in the world on a two-wheeled electric vehicle in 1973, reaching a speed of 275 km per hour (171 mph) on one of his custom-built electric racing bikes. He is now in business with his son, and together they have created a unique example of personal transportation, a registerable street vehicle for congested, smoggy areas. The 'Sparrow' is designed to take up the space of a motorbike and is licensed as such. Ninety-five percent of car journeys are made by only one person. The car is single seated and therefore compact, four being able to park in the space taken up by one average-sized family vehicle. It is cheap, retailing for only $12,500, and is ideal as a second car for commuting. A station car. Corbin Motors envisage mass-transit systems buying a fleet of Sparrows and hiring them on a weekly basis to commuters. They would come with a guaranteed parking place at either end of the journey, complete with re-charging points as well as a charger for the home garage. Once in the city they can use commuter lanes thus leaving congestion behind and, due to zero emissions, will not add to the pollution crisis. The organic shape of the chassis is not only aerodynamic but also a visual signifier of this little bird's desire to improve the environment.

'I must commit myself to reforming the environment and not man; being absolutely confident that if you give many the right environment he will behave favourably.' R. Buckminster Fuller

Above: No Picnic. Lounger. Polyurethane foam, textile, steel; h. 54cm (21^{1}4in) w. 100cm (39^{3}8in) d. 88cm (34^{5}8in). Felicerossi, Italy

Opposite: Ross Lovegrove. Armchair, Lovenet. Galvanized and powdered steel, PE, stainless steel; h. 90cm (35^{5}8in) w. 150cm (59in) d.80cm (31^{1}2in). moooi©, The Netherlands

Above: Patricia Urquiola. Armchair, Fjord. Steel, flame-retardant polyurethane foam; armchair: h. 102cm (40^{1}8in) w. 95cm (37^{3}8in) d. 80cm (31^{1}2in). Moroso SpA, Italy

Reminiscent of the 'Egg Chair' by Arne Jacobsen, Urquiola has updated the Scandinavian tradition of harmonizing with the natural world by using new materials and production techniques. The asymmetrical outline of the Fjord chair takes on the form of a relaxing body structure, while also recalling the Nordic lands of cracked rocks smoothed and shaped by the sea, after which they have been named.

Opposite: Stefano Giovannoni. Floorstanding WC. Ceramic; h. 46cm (18^{1}8in) w. 58.5cm (23in) l. 39cm (15^{3}8in). Alessi SpA, Italy

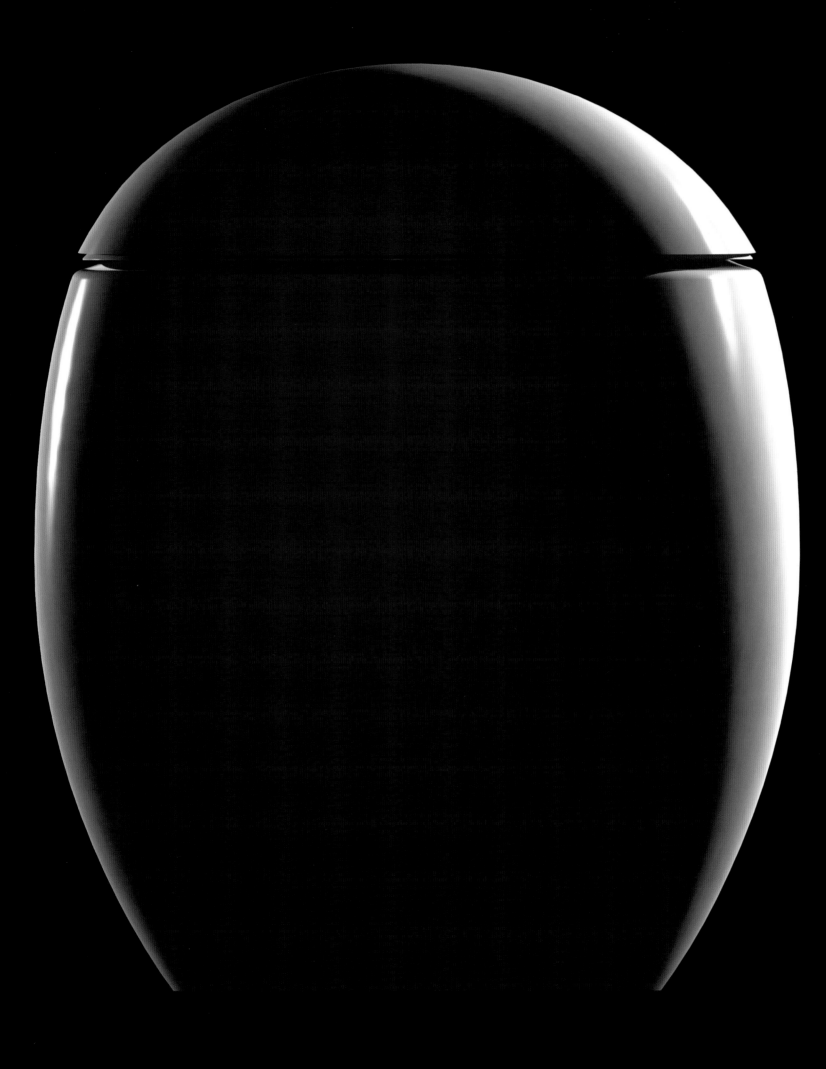

Top: Enrico Azzimonti and Jordi Pigem. Bed, Orms. Wood, polyether, fabric, stainless steel; h. 45cm (17³⁄4in) w. 100cm (39³⁄8in) l. 275cm (108¹⁄4in). Prototype

Above: Philippe Starck. Armchair, Zbork. Polyethylene. Kartell SpA, Italy

Opposite: Ross Lovegrove. Bench, bdlove. Rotation-moulded polyethylene; h. 94cm (37in) w. 130cm (51¹⁄8in) l. 265cm (104³⁄8in). Bd Ediciones de Diseño, Barcelona

Above: Luca Bonato. Armchair, Sections. Acrylic; h. 108cm (42½in) w. 120cm (47¼in) d. 86cm (33⅞in). Fusina srl, Italy

Luca Bonato's chair is inspired by CAD-CAM drawings and is fabricated in laser-cut clear acrylic sheets.

Opposite above: Mauro Mori. The Head. Albizia wood; h. 95cm (37⅜in) w. 55cm (21⅝in) d. 53cm (20⅞in). One-off

Opposite below: Carl Öjerstam. Lounger, Storvik. Clear lacquered rattan; h. 77cm (30⅜in) w. 105cm (41⅜in) l. 130cm (51⅛in). IKEA, Sweden

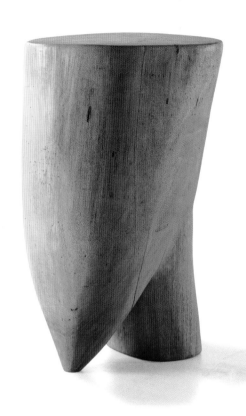

Opposite: Scot Laughton. Hanging lamp, Jube Jube. Ceramic, stainless steel; h. 12cm (4³⁄4in) w. 19cm (7¹⁄2in) l. 42cm (16¹⁄2in). Lolah, Canada

Below: Von Robinson. Play Orbiter. Moulded glass reinforced resin; h. 83.8cm (33in) d. 91.4cm (36in) l. 83.8cm (33in). VRiD, USA

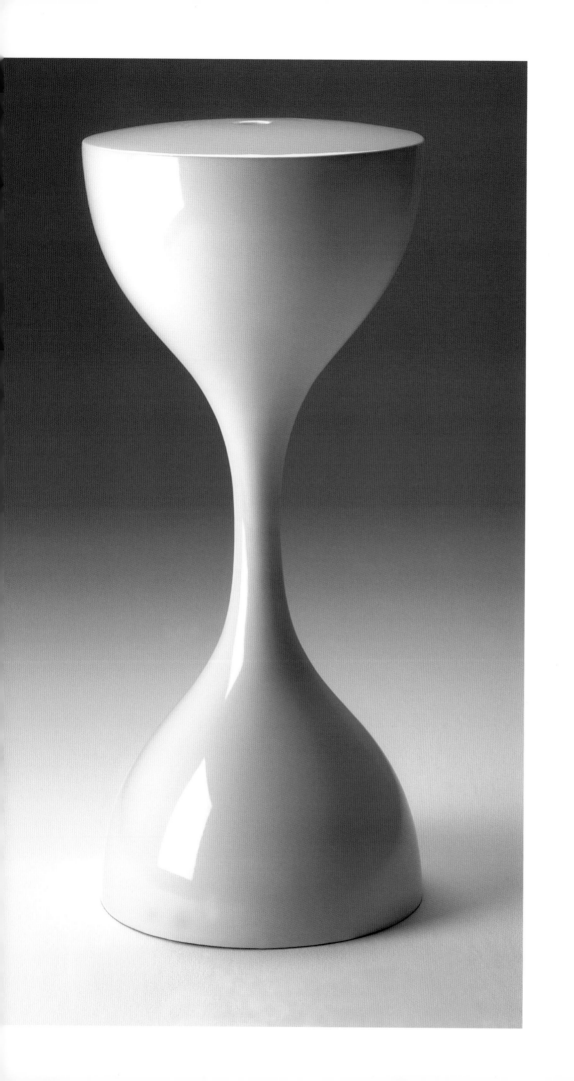

Opposite: Christopher Streng. Bar stool, Belle. Fibreglass: h. 76.2cm (30in) w. 35.5cm (14in) diam. 35.5cm (14in).
Christopher Streng Inc, USA

Above: Claudio Colucci. Lamp, Squeeze. Corian®; h.178cm (70in) diam. 47cm (18¹⁄2in). Limited batch production;
Créa Diffusion, France

Opposite: Christiane Müller and Liset van der Scheer. Rug, Salto. Pure new wool; h. 4.5cm ($1^3$4in). Danskina, The Netherlands

Top: Fernando and Humberto Campana. Sofa, Boa. Polyurethane foam, velvet; h. 80cm ($31^1$2in) w. 150cm (59in) l. 270cm (106^{1}4in). Edra SpA, Italy

Above: Guido M. Rosati. Armchair and sofas, Papillon. Steel, undeformable polyurethane, dacron, protective tissue; armchair: h. 70cm (27^{1}2in) w. 100cm (39^{3}8in) d. 100cm (39^{3}8in), sofas: h. 70cm (27^{1}2in) w. 150/200cm (59/78^{3}4in) d. 100cm (39^{3}8in). Giovannetti srl, Italy

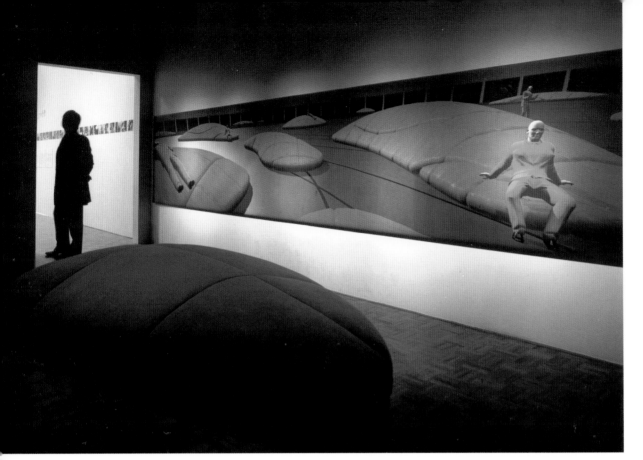

Wim Poppinga. Green Rolling Hills. Auping Foundation, The Netherlands

For last year's Milan Furniture Fair, the Auping Foundation showcased an exhibition of designs entitled 'Mobile Dreaming – designing beyond the bedroom'. It was somewhat off the beaten track, but well worth the visit for the amusing and often irreverent way that selected designers had handled the concept that in the modern western world it cannot be taken for granted that people always enjoy their night's rest in their own bedroom. In our mobile society, traditional modes of living are altering, from where and how we eat to nomadic working. Eibert Draisma's 'Nightcap' (overleaf) is an effective construction to stop a head nodding when falling asleep on the bus or train. The hat supports the head in all directions but is slightly elastic so that the rocking motion of the head is not interrupted. 'Green Rolling Hills' by Wim Poppinga also addresses the need for the natural in an increasingly industrialized and mechanized world. Christopher Seyferth's 'Domestic Landscape' is the artificial rural idyll that city-dwellers wish for in the middle of the working day. Designed as street furniture, the undulating hills are covered in a soft version of Astroturf and offer a dreamscape for the harassed. Seyforth is proposing a piece of non-design that aims to adapt to the activities or inactivities of the user. It replaces our idea of traditional furniture as it moulds itself to human contours and is a comforting organic and ever-changing structure. The landscape can be shaped and transformed into a table, a sofa, a chaise, or whatever.

Above: Eibert Draisma. Nightcap. Auping Foundation, The Netherlands

Opposite: Monkey Boys. Lamp, Tent. Fabric, fibreglass; h. 80cm (31^{1}2in) w. 60cm (23^{5}8in) l. 110cm (43in). Prototype

Overleaf left: Angelica Gustafsson. Tumblers, Bamboo. Handmade glass; h. 16cm (6^{1}4in). Skruf, Sweden

Overleaf right: Mark Dyson and Monika Piatkowski. Textile, Circulation. Wool felt pellets. Hive, UK

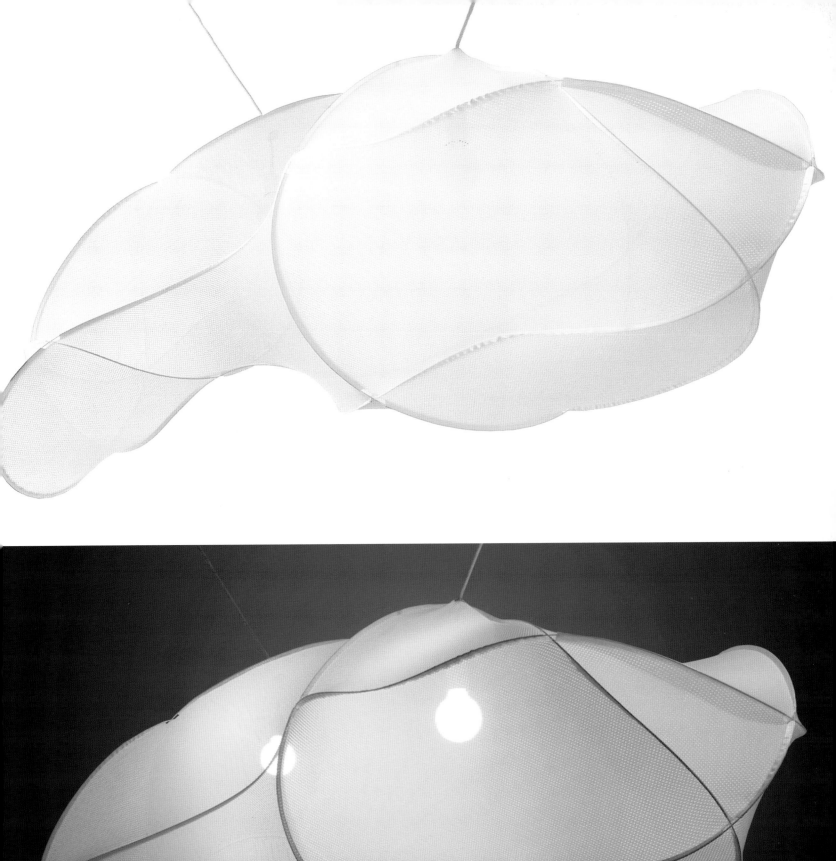

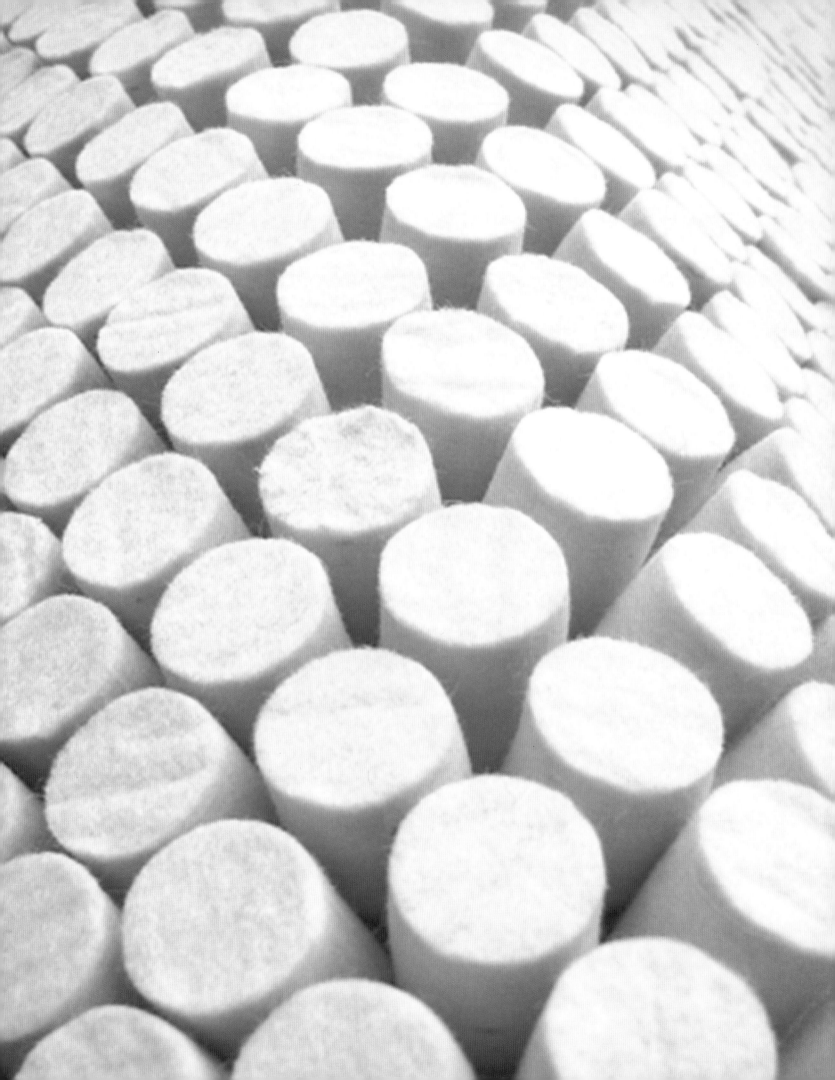

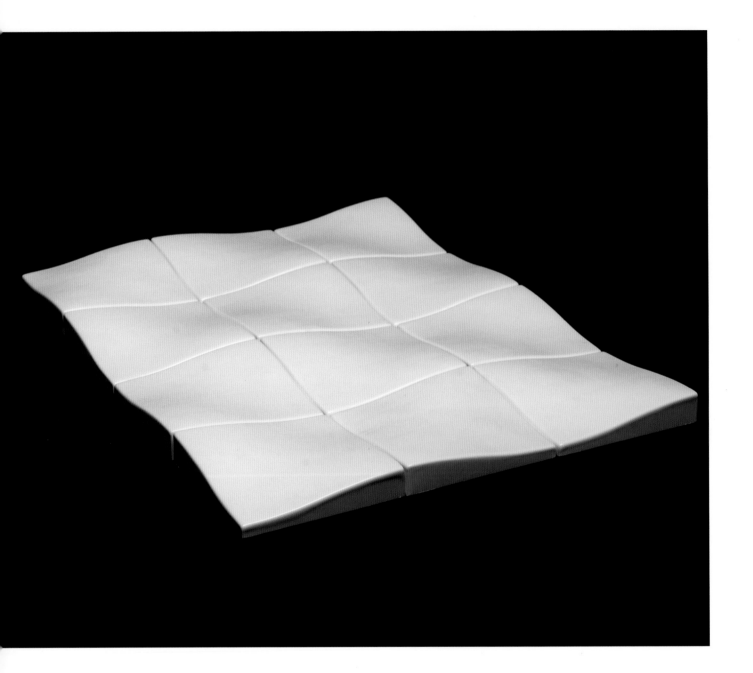

Nick Rennie. Wavy tile. Ceramic; h. 9cm (3½in) w. 50cm (19⅝in) l. 50cm (19⅝in). Happy Finish Design, Australia

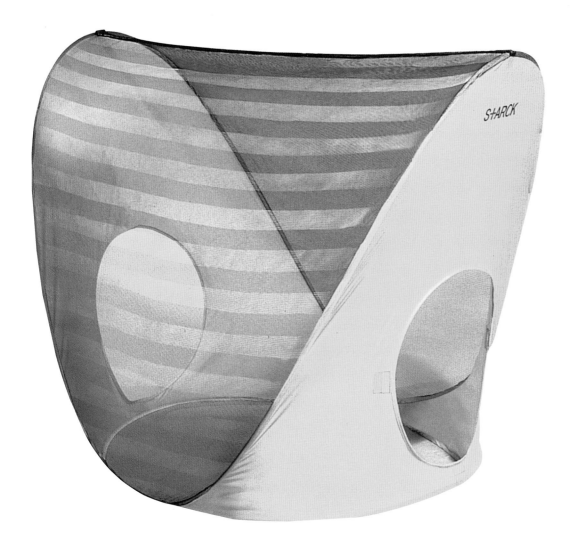

Karim Rashid's ambition is to democratize design by bringing it to the shopping malls of middle America. He is one of the leading players in the democratic design wave, which has grown to mirror the interest the man in the street has developed for good products at reasonable prices that have the added bonus of having been designed by someone they've read about in the style press. A leading exponent of this ideology is one of Karim's mentors, Philippe Starck. As a young designer, Rashid idolized Starck for his diversity and his extravagant and extroverted opinions. This admiration has not waned, but as his own career has developed has been adopted and personalized in his own philosophy and vision. Both are prolific and are concerned with bringing their enthusiasm to the general public. Starck has just completed a range of products for Target, one of the largest supermarket chains in the US. Each outlet, the size of several football pitches, is packed full of mass-produced banal merchandise, but now taking pride of place at the end of most aisles are advertisements for the various products Starck was recently commissioned to create. Starck maintains that he is as happy designing an inexpensive item for all as he is an over-priced project for the rich and famous. So as a balance to his super chic and luxurious hotel interiors for Schrager, his one-off technically advanced yacht and the crystal lights 'Cicatrices des Lux' for Flos, to name just a few, he has now developed 'Starck Reality'. This line comprises functional yet extremely beautiful baby products, kitchen items, toiletries, furniture and stationary that retail from between $2 and $40. Starck says, 'For twenty years I have been trying to show that the real elegance is in the multiplication'. He believes design should be accessible to all and should enhance, rather than detract, from the quality of life. 'Starck Reality' was inspired by watching the trouble his wife had retaining her elegance while caring for a young child, being forced to use necessary but often tackily designed baby paraphernalia. By setting the example of bringing quality items to the general public, Starck hopes that others will follow. IKEA were pioneers in bringing contemporary design to the masses, but as more and more manufacturers realize that there is a growing market for this kind of inexpensive product others will need to compete. As Starck points out, 'This means good objects of good quality everywhere at the right price. And that's what interest's me.'

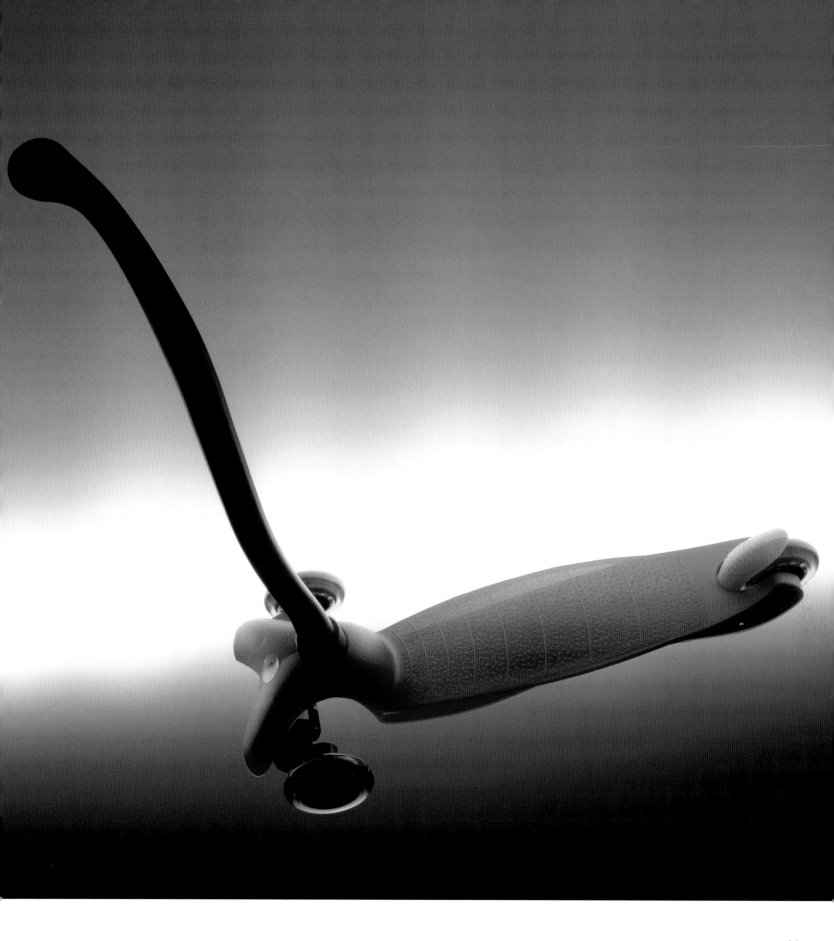

Opposite: Mauro Mori. Lady Mamba. Black Belgium marble; h. 112cm (44in) w. 35cm (13^{3}4in) l. 26cm (10^{1}4in). One-off

Above: Juergen R. Schmid. Scooter, Bobbi-board PA6 GF30. H. 68cm (26^{3}4in) w. 22cm (8^{5}8in) l. 55cm (21^{5}8in). Micro Mobility Systems D GmbH, Germany

There is a human need for the decorative: style, language and surface treatment that connote something deeper or more vocal than purity of colour and form. I love decoration – when it has meaning. Historically decoration always had meaning and status that made reference to religious, moral, gentrified or civil communication. With mass-production it became a luxury, so the machine age parted with the ornate. Then technology reached a point where decoration became more cost effective, a way of hiding flaws, of camouflaging defects, and of 'cheapening production'. Ornamentation no longer means more work or cost as it once did with the handmade or hand applied. Some crystal today is highly 'cut' or decorated to hide air bubbles, fake wicker-weave holes in plastic hampers reduce the amount of plastic or hide sink marks and bad quality mouldings. Simply surfacing the product has become the fastest and simplest way of offering consumer choice and tribe-niche. Decoration is now an opportunity to diversify and customize. New meanings have surfaced. Signs for technology such as 'fake venting' (originally real vents), Op art, Pop art, illusory perception and graffiti are all ways to personalize or rarify mass-produced goods. In our digital age an 'infostetic' decorative language is emerging, a techno-rave new world of images, symbols, argots, textures and 'digipop' vernacular.

EMBELLISHMENT

'The anti-decorative dictate ... is a modernist mantra if ever there was one ... But maybe times have changed again; maybe we are in a moment when distinctions between practices might be reclaimed and remade – without the ideological baggage of purity and propriety attached.' Hal Foster

Previous page: Karim Rashid. Fabric, Digiweave. Edra SpA, Italy

Above: Christian Ghion. Carpet tile, Collection Privee 3/sub-version. W. 50cm (19⁵⁄8in) l. 50cm (19⁵⁄8in). Tarkett Sommer, France

Yoshiki Hishinuma. Textile, No. 9. Hishinuma Associates, Japan

Opposite: Teresa Sapey. Table, Tavolo Lei. Steel, glass; h. 72cm (28^{3}8in) diam. 40cm (15^{3}4in). Prototype; Edizione Straordinaria Com Varese, Italy

Below: Yoshiki Hishinuma. Textile, No. 11. Hishinuma Associates, Japan

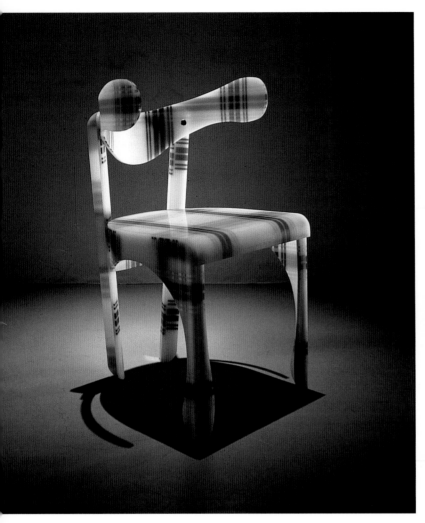

Tsutomu Kurokawa. Chair, Mapell. Polycarbonate, acrylic resin; h. 84cm (33in) w. 42cm (16$\frac{1}{2}$in) d. 51cm (20in). waazwiz Ltd, Japan

Tsutomu Kurokawa. Chair, Qunte. Acrylic resin; h. 80cm (31$\frac{1}{2}$in) w. 47cm (18$\frac{1}{2}$in) d. 45cm (17$\frac{3}{4}$in). waazwiz Ltd, Japan

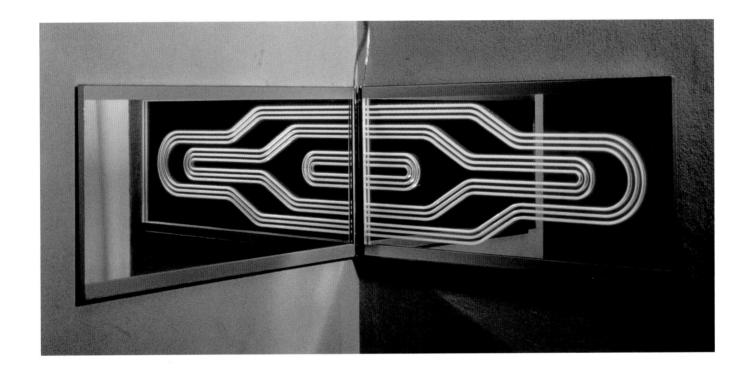

Stefano Miceli. Lamp, V-Doub 2. Aluminium, acrylic, LED SMD; h. 30cm (17^{7}8in) w. 60cm (23^{5}8in) l. 60–120cm (23^{7}8–47^{1}4in).
Limited batch production; Memphis srl and Post Design, Italy

Top: Versace. Sofa, Wellington, with printed cushions, Floral Leopard. Gianni Versace Home Collection, Italy

Above: Paul Smith. Armchair, Mondo. Wood, polyurethane foam, macroter lacquer, fabric; h. 72cm (28^{3}8in) w. 73cm (28^{3}4in) d. 69cm (27^{1}8in). Cappellini SpA, Italy

Opposite: Paul Smith. Cabinet, Mondo. Matt lacquered wood, silk screen; h. 108cm (42^{1}2in), Cappellini SpA, Italy

Paul Smith's tailoring is 'classic with a twist', typically top quality suits with a sense of humour. All his designs have hidden accents, something unexpected – a Union Jack emblazoned on the lining of a dark blue jacket, a coloured print tucked inside the cuff of a classic striped businessman's shirt. It is this tongue-in-cheek quality that prompted Giulio Cappellini to ask him to develop a line of furniture for Mondo. Starting from some basic products already being manufactured by Cappellini, the brief was to create a collection that would represent the bourgeois world in an ironic way by adding a touch of 'new baroque'. Design is becoming ever more cross-disciplinary. When interviewed, Paul Smith is as much at home talking about graphics, art and architecture as about fashion. The title of his new book 'You Can Find Inspiration in Everything, and If You Can't, Look Again' says it all. Yet when he was initially approached by Cappellini, Smith told me that he was very unsure as he has many friends who are well-known for their product and furniture design: 'I felt that it might be disrespectful to enter their world as I have no formal training in furniture design. When I finally accepted the commission I thought that my approach should be more as a stylist rather than a designer.' However, he did not resort to covering sofas in a typical Paul Smith striped fabric, but together with Cappellini has produced a global concept that strongly identifies the collection and comes complete with his signature surprises and trompe l'oeil effects. It's styling with a statement, and not embellishment for embellishment's sake.

Above left: Emmanuel Babled. Vase, Overcross, and plate, Overflower. Handblown glass; vase: h. 43cm (16^{7}8in) diam. 40cm (15^{3}4in), plate: diam. 61cm (24in). Limited batch production; Idem, Italy

Above right: Guido Venturini. Large vase, Lingam. Ceramic; h. 138cm (54^{3}8in). Alessi SpA, Italy

Opposite: Emmanuel Babled. Hypnos range. Crystal and Corian®. Baccarat, France

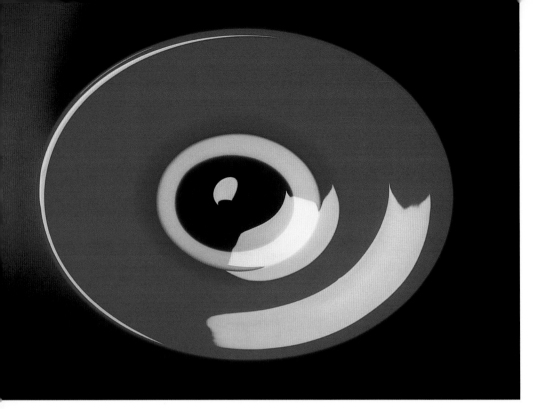

Top: Carlo Moretti. Plate, Tramonto in Laguna. Murano crystal; h. 6.5cm (2¹2in) diam. 35.5cm (14in). Limited batch production; Carlo Moretti srl, Italy

Above: Matt Sindall. Dining/conference table. Satinated steel, metacrilate; h. 79cm (31¹8in) w. 200 cm (78³4in) d. 100cm (39³8in). Sawaya & Moroni SpA, Italy

Opposite: Ross McBride. Wall panel, Ripple. Fireproof ABS; h.3cm (1¹8in) w. 40cm (15³4in) l. 40cm (15³4in). David Design, Sweden

136

Fernando and Humberto Campana. Chair, Sushi. Plastic, felt, fabric, carpeting. Edra SpA, Italy

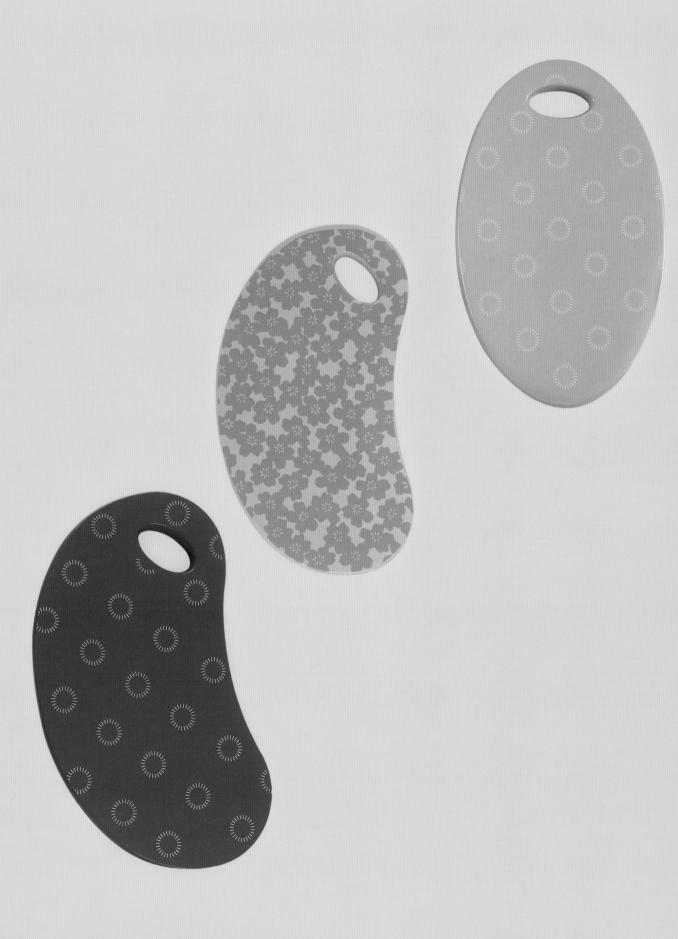

Opposite: Michael Graves. Garden kneeling pads. Foam; w. 28.6cm (11$\frac{1}{8}$in) l. 47cm (18$\frac{1}{2}$in) d. 3.1cm (1$\frac{1}{4}$in). Rumford Gardner, USA

Above: Toshiba. Digital camera, T10. Toshiba, Japan

The desire to do more with a single object is an intriguing design objective. Since the ancient Egyptians we have witnessed how any given product may be endowed with multiple functions or can be constructed with module components or 'building blocks'. A bed that becomes a divan that becomes a chaise longue answers to a desire for more, just like James Bond's Alfa Romeo transforming into a plane and then an underwater SUV. There are pragmatic reasons, such as small urban spaces, where peoples' needs for multi-functionality is obvious. But the natural poetic articulations of surprise, of the phenomenological change over time, is what humans also embrace. A typical question raised by this is whether a product loses some of its performance, so you end up compromising quality. Is it better to have three objects that perform really well, rather than one that performs the three functions badly? Do we actually spend the time changing the object or does it always remain in just one of its forms? Is it merely a marketing seduction? We have several orders of real change and variation in our built landscape – phenomena, modularity, serialization, decoration, diversity and individualism. The notion of reconfiguration or modularity contributes to a behavioural vicissitude. One of the most multi-functional icons of the last century is the wood crate, a building block that started for milk storage, then became a vinyl record holder, then shelving, then bed supports, and so on. Generations adapted the ubiquitous object to their personal use. Interpretation of the object is important here. A module allows for customization ad infinitum, and reconfigurability engages us physically and mentally.

Multiplicity

'There is the right size for every idea.'
Henry Moore

Previous page: Karim Rashid. Chair, Kapsule Kids. Injection-moulded polypropylene; h. 44.4cm (17$\frac{1}{2}$in) w. 46.7cm (18$\frac{3}{8}$in) l. 54.8cm (21$\frac{1}{2}$in). Bozart, USA

Above: David Quan. Stool, Storit. Foam, plastic; h. 52cm (20$\frac{1}{2}$in) diam. 33.6cm (13$\frac{1}{4}$in). Umbra, USA

Patricia Urquiola. Containers, Fat-Fat and Lady-Fat. Lathe-turned metal, nickeled or varnished; Lady-Fat: h. 30cm (11⁷⁄₈in) diam. 116cm (45⁵⁄₈in) or h. 35cm (13³⁄₄in) diam. 85cm (33¹⁄₂in), Fat-Fat: 45cm (17³⁄₄in) diam. 66cm (26in). B&B Italia, Italy

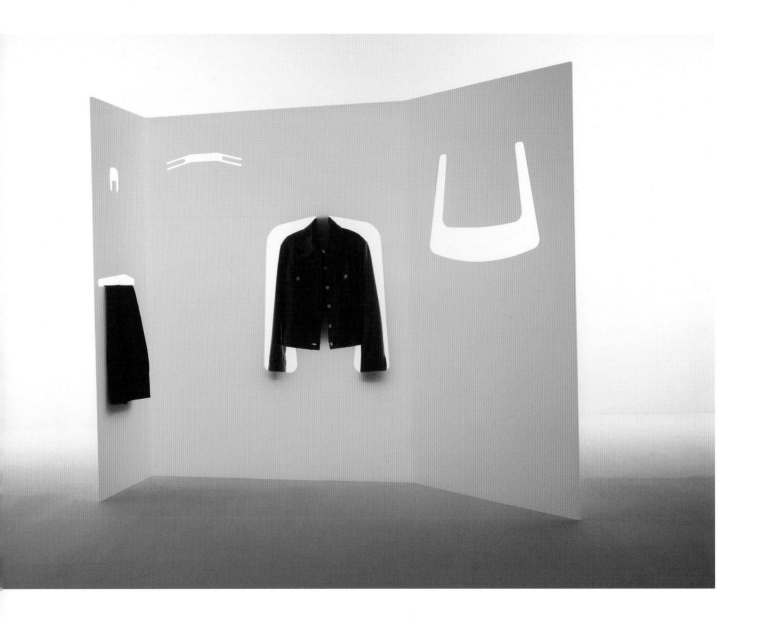

Above: Barber Osgerby. Clothes hanger/screen, Stencil. Acrylic; h. 170cm (67in) w. 1.2cm (^{1}2in) l. 272cm (107in). Cappellini SpA, Italy

Barber Osgerby's 'Stencil' is a perfect reduction of form and function. It can be used as a room divider with abstract forms cut out haphazardly or they can be engineered to hold pieces of clothing, transforming the piece into a functional nightstand or wardrobe for an open plan home or hotel room.

Opposite: Lars Jorge Diederichsen and Fabiola Bergamo. Chair, Giro. Polypropylene, stainless steel; h. 84cm (33in) w. 67cm (26^{3}8in) d. 50cm (19^{5}8in). Terra Design, Brazil

'Giro' is suitable for internal or external use. The chair can be refolded to form a seat with or without armrests.

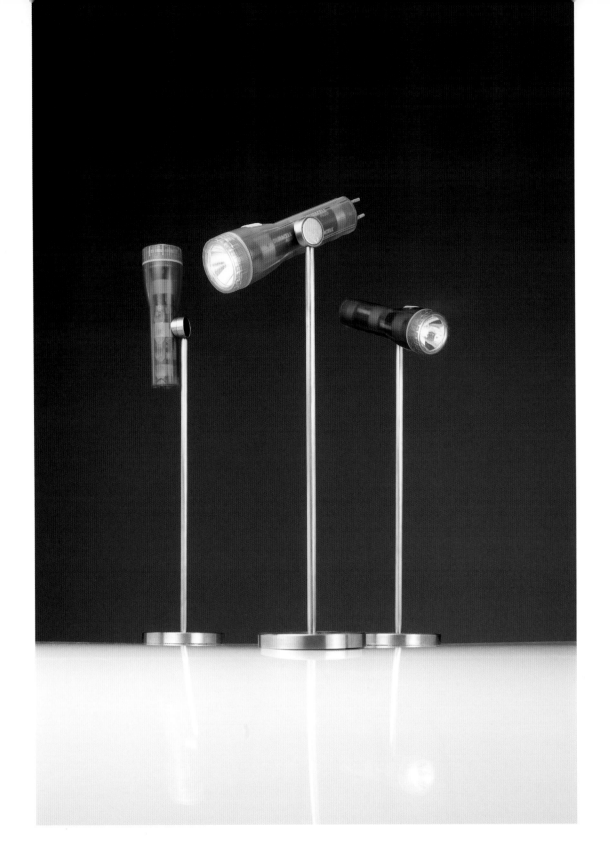

Lorenzo Damiani. Lamp, Black Out. Steel, plastic; h. 38cm (15in) diam. 12cm (4³⁄₄in). Prototype

Since the stereo and CD stand 'Flex', Lorenzo Damiani has built up a reputation for deceptively simple yet ingenious designs. In his lamp 'Black Out', the plastic light diffuser is a torch connected to a stem by means of a magnet. It can therefore rotate 360 degrees and can be removed. The light source works from two rechargeable batteries contained in the lamp itself, which has a built-in plug that can be inserted into common sockets. The result is a new typology, making a virtue out of an item that is normally stored in a drawer until needed. Now why didn't I think of that!

Patrick Chia. Lamp/stool, Ufosausoo. Opalescent plexiglas, chromed metal, fluorescent bulb; h. 79cm (31in) d. 40cm (15$\frac{3}{4}$in) w. 38cm (15in), h. 115cm (45$\frac{1}{4}$in) d. 40cm (15$\frac{3}{4}$in) w. 48cm (18$\frac{7}{8}$in). BRF srl, Italy

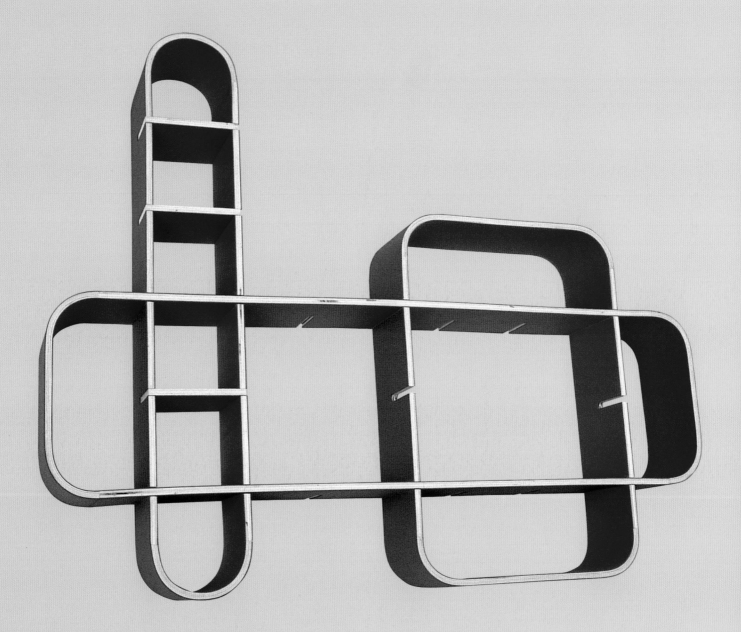

Viktor Jondal. Shelf unit system, Mr Tall, Big & Slim. Moulded plywood, rubber laminate; h. 168cm (43in) w. 138cm (55in) d. 20cm (7¹2in). Limited batch production

Scot Laughton. Stool, Situ. Fibreglass, moulded polyurethane, mould colourfast surface colouring; h. 43.2cm (17in) w. 45.7cm (18in) l. 82.5cm (32$\frac{1}{2}$in). Lolah, Canada

'Situ' can be used as a table-height stool, a leaning post, a perch or a resting place.

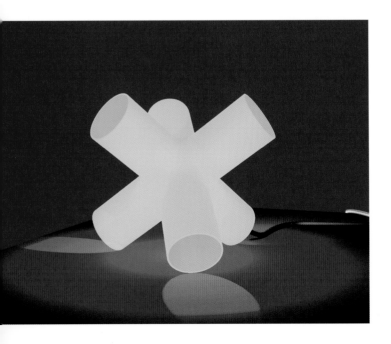

Opposite above and below left: Jan Melis and Ben Oostrum. Table or pendant lamp, Crosslight. H. 38cm (15in) w. 38cm (15in) l. 38cm (15in). MNO, The Netherlands

Opposite below right: Jacopo De Carlo, Andrea Gualla and Raffaella Godi. Mirror/wall decoration/lamp, Merlino. H. 62cm (24^{3}8in) w. 62cm (24^{3}8in) d. 12cm (4^{3}4in). DeCarloGualla, Italy

Merlino solves the problem of mirror, picture and lamp competing for space on the same wall as it acts as all three. A random sequence of the different luminous sections gradually reveals the different functions. The reflecting surface of the mirror transforms into a printed image, which in turn fades when backlit and turns into a lamp.

Above: Inga Sempé. Lamp, PO/0203. Varnished metal; h. 70cm (27^{1}2in) w. 50 cm (19^{5}8in). Cappellini SpA, Italy

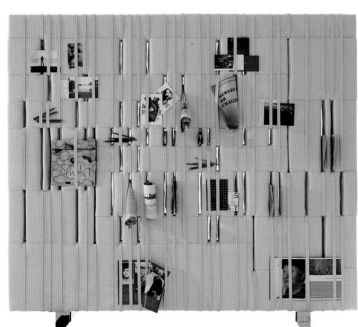

Opposite: Carsten Gerhards and Andreas Glücker. Soft Wall. Felt, chrome; h. 207cm (81$\frac{1}{2}$in) w. 20cm (7$\frac{3}{4}$) l. 250cm (98$\frac{3}{8}$in). B&B Italia, Italy

Above: Piero Lissoni. Sofabed, Reef. Steel, polyurethane foam, polyester, fabric; h. 69cm (27$\frac{1}{8}$in) w. 200cm (78$\frac{3}{4}$in) d. 84.5cm (33$\frac{1}{4}$in). Cassina SpA, Italy

Above: Sean Yoo. Seating system, Flip. Chair: h. 77cm (30^3$_8$in) w. 57cm (22^3$_8$in) d. 108cm (42^1$_2$in), chaise longue: h. 62cm (24^3$_8$in) w. 195cm (76^3$_4$in) d. 65cm (25^5$_8$in). Apt5 Design, Italy

Opposite: Johannes Fuchs. Sofabed. Steel tube, upholstery; h. 73cm (28^3$_4$in) w. 90cm (35^3$_8$in) l. 210cm (82^5$_8$in). One-off; Johannes Fuchs Produkt Design, Germany

Right: Marc Krusin, Hammock/deckchair, Crossed. Steel tube, PVC sheet: h. 60cm (23⅝in) w. 100cm (39⅜in) l. 120cm (47¼in). Marc Krusin, Italy

'Poetry in motion' – Krusin's 'Crossed' is structure pared down to the minimum. It is constructed from two tubes placed one inside the other and held in place by the PVC seating above and two tension cables below. There are no screws, hinges, guides or constraints of any kind, so the lounger is always moving. The seating position is controlled only by the weight of the user and the position in which they sit. Halfway between a hammock and a deckchair, you can sit up or lie down in the knowledge that all that is holding you there is gravity. Krusin's enquiring, mathematical mind focuses on construction and material. His designs do not have derivatives, but are the result of distancing himself from traditional uses and historical visual restraints to concentrate on what is suggested to him by the materials he is using, letting the result define the aesthetic.

Opposite above and centre: Arik Levy. Sofabed, Arik. Polyurethane foam, steel, feather-filled cushions, fabric; h. 86cm (33⅞in) w. 245cm (96½in) d. 120/150cm (47¼/59in). Ligne Roset, France

Opposite below: Eric Jourdan. Chaise longue, Tolozan. Steel, thermoformed polystyrene, polyethylene, wood veneer; h. 85cm (33½in) w. 76.2cm (30in) d. 137.8cm (54¼in). Ligne Roset, France

Opposite above: Walter Craven. Side table, Bunny. Thermoformed plastic, tubular steel, upholstered cushions; h. 40.6cm (16in) w. 30.5cm (12in) l. 40.6cm (16in). Walter Craven Design, USA

Opposite below: Konstantin Grcic. Side tables, Diana. Powder-coated sheet metal; h. 42cm ($16^1/2$in) w. 53cm ($20^7/8$in) d. 25cm ($9^7/8$in). ClassiCon GmbH, Germany

Above: Marcel Wanders. Low stools, Pebbles. Plastics; h. 37cm ($14^1/2$in) w. 42cm ($16^1/2$in) d. 42cm ($16^1/2$in). Magis srl, Italy

Above: Francesco Binfaré. Seating system, Damier. Edra SpA, Italy

Opposite: Denis Santachiara. Sofa, Giubbe Rosse. Plastic, fabric; h. 70cm (27$\frac{1}{2}$in) w. 192cm (75$\frac{1}{2}$in) d. 94cm (37in).
Styling srl, Italy

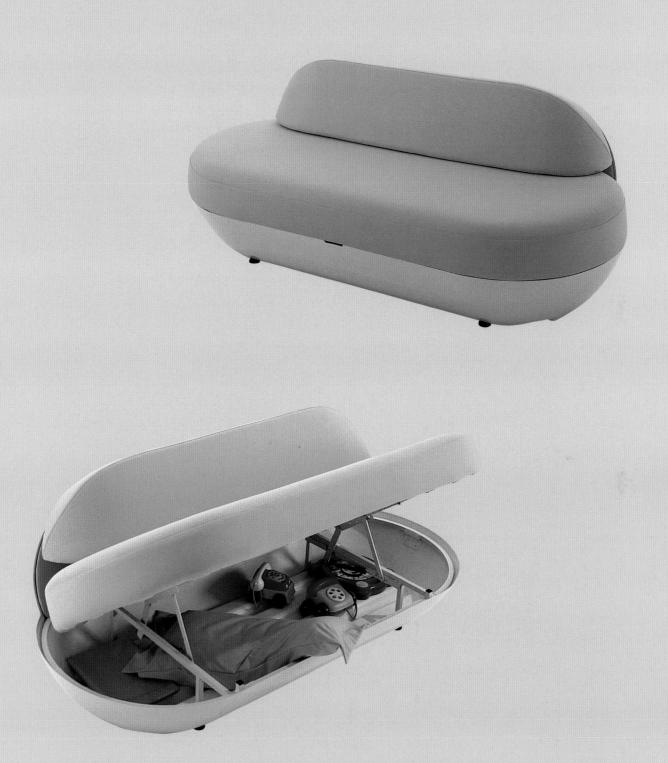

Diego Fortunato. Magnetic cushions, Cojines! Cojines! Technological fabric, magnet; h. 15cm (5^{7}8in) w. 48cm (18^{7}8in) l. 48cm (18^{7}8in). Nani Marquina, Spain

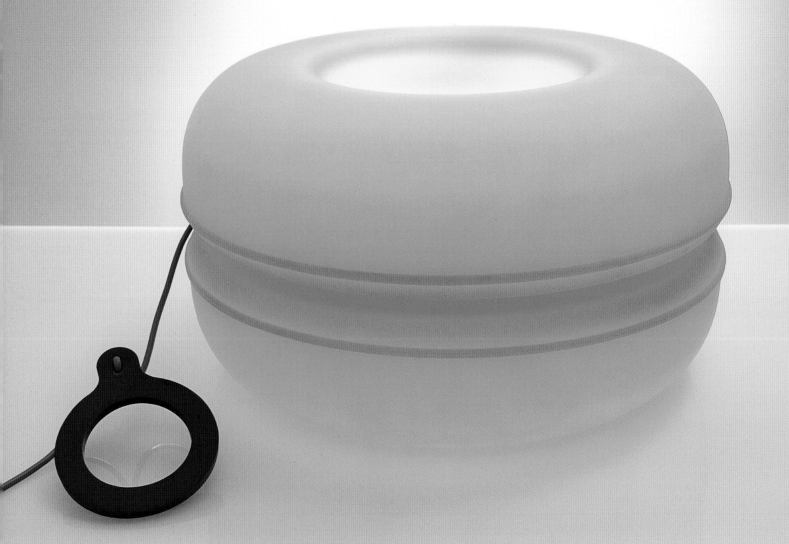

Above: Celina Clarke, Simon Christopher and Paul Angus. Lamp, Yo. Polyethylene; d. 34.5cm (13^{1}2in) diam. 60cm (23^{5}8in).
ISM Objects, Australia

Opposite: Izumi Kohama and Xavier Moulin. Homewear stoolpants. Inflatable PVC, plastic tubing. ixilab, Japan

Opposite: Gerhart Ploegstra. Dancing Chair. Stretch fabric, iron; h. 60-100cm (23$\frac{5}{8}$–39$\frac{3}{8}$in) w. 80cm (31$\frac{1}{2}$in) l. 80cm (31$\frac{1}{2}$in). Univorm, The Netherlands

Right: Christophe Francois. Chaise longue/meridienne/sofa, Campos. Wenge, light beech, fabric; w. 200cm (78$\frac{3}{4}$in) d. 80cm (31$\frac{1}{2}$in). Kyo Design, France

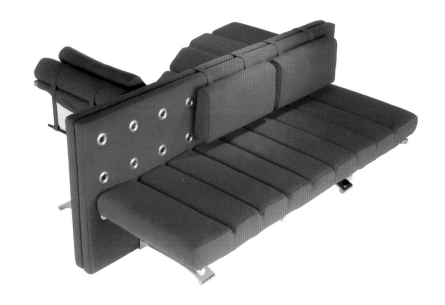

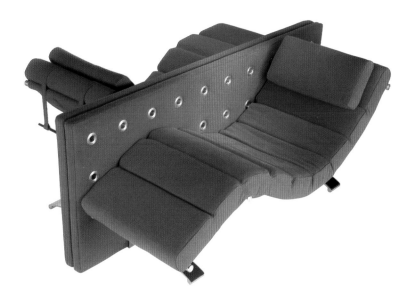

Dene Happell. Tile, Wedge. Ceramic; h. 15cm (5⁷⁄8in) w. 15cm (5⁷⁄8in) d. 40cm (15³⁄4in). Happell, UK, and Lolah, Canada

Cordula Kafka. Indoor and outdoor seat, Blast. Laminated fibreglass, leather cushion; h. 25cm (9⁷⁸in) w. 80cm (31¹2in) l. 80cm (31¹2in). Prototype

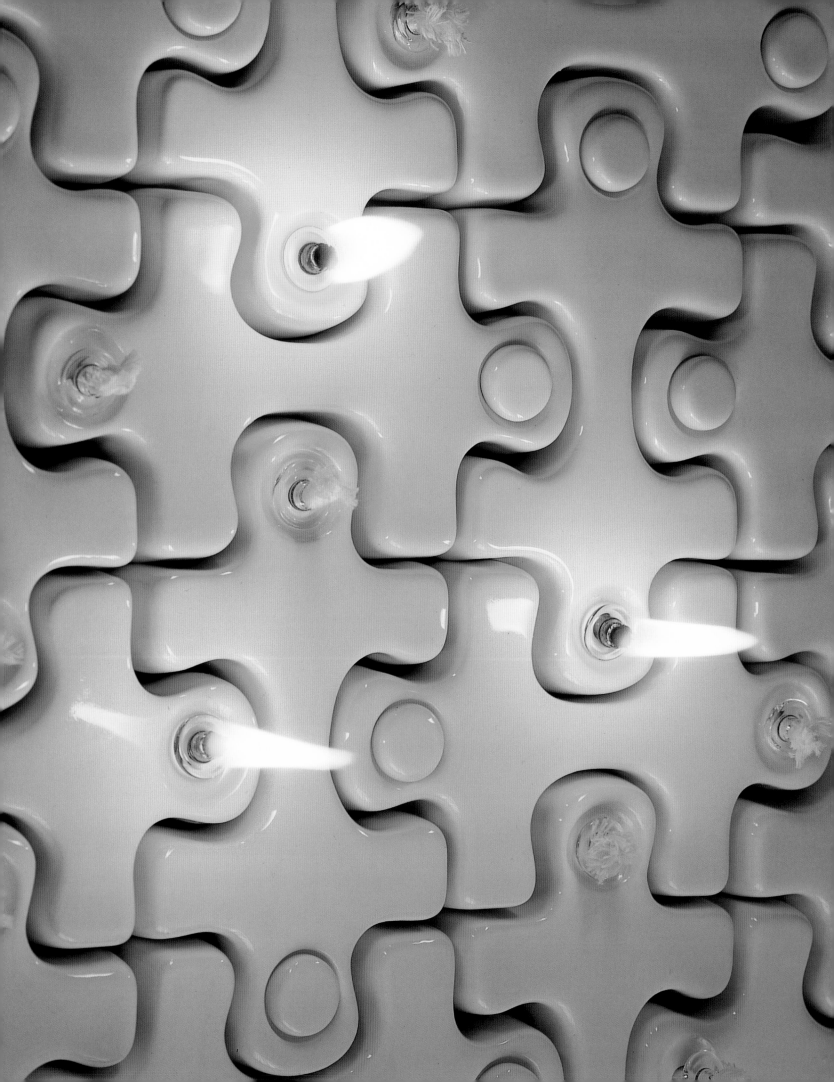

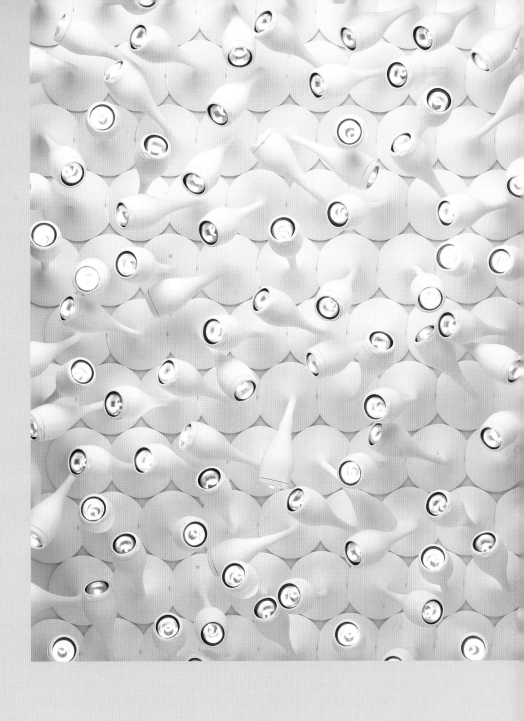

Opposite: K.C. Lo. Oil burner, Jigsaw. Ceramic, glass, cotton; h. 3.5cm (1^{3}8in) w. 12.5cm (4^{7}8in) l. 7cm (2^{3}4in).
Limited batch production

Above: Carl Öjerstam. Wall spotlight, Flamma. Polypropylene, synthetic rubber; diam. 8cm (3^{1}8in). IKEA, Sweden

'Maximum simplicity and maximum limitation in the use of time and material' was the bausian ideal. Modernism eradicated the idea of extraneous decoration, so that a product spoke only of its necessary inseparable constructs. Design was based on the most reduced structure to achieve a 'pure' minimized condition. The result was a landscape of clean, unadulterated aesthetics, where abstraction consisted of geometry and mass-production to create a universal 'global' result. The perfect composition of the 'essential'. Minimalism lacks the human aspect due to its sacred geometric language that is removed from history and nature. This is possibly why it is seen as a divine spiritual aesthetic – non-referential, purely abstract and artificial. I like design that traverses the boundaries of the associative and touches the sensual, that goes beyond the modernist doctrine of *belnahe Nichts* (almost nothing). Minimalism seems to be shifting to a more sensual minimalism or 'sensualism', where objects communicate, engage and inspire, yet remain fairly minimal a posteriori. They can speak simply and directly, without being superfluous. These objects can be smart, fuzzy or low-tech, while still being informed by these factors. The marriages of new materials and technology with pure geometry each carries significance, symbolic of a personal language. The new materials are reflective, smooth, glowing, pristine, glossy, mutable, flexible, tactile, synthetic, high contrast, sensual, cold and hot, chrome, liquid, glass, lacquered, fluorescent, smart, interactive and illusory. This visual language is undeniably beautiful, but is it human? Is it a philosophy that goes beyond the design community and touches peoples' souls? Regardless, it is here to stay and to reach even higher forms of perfection and beatitude.

MINIMUM

'The devil hath power to assume a pleasing shape.' William Shakespeare

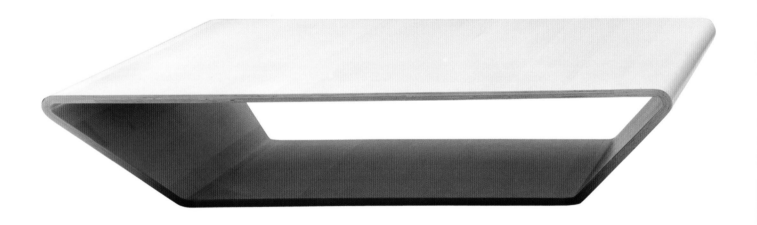

Previous page: Karim Rashid. Issey Miyake Men's 2-in-1 Fragrance Set. Polyethylene; h. 8cm (3$\frac{1}{8}$in) l. 11cm (4$\frac{3}{8}$in) d. 3cm (1$\frac{1}{8}$in). Issey Miyake Perfumes, France

Top: Ferruccio Laviani. Armchair, Roma. Plywood; h. 72cm (28$\frac{3}{8}$in) w. 49cm (19$\frac{1}{4}$in) l. 62cm (24$\frac{3}{8}$in). Emmemobili, Italy

Above: Eero Koivisto and Ola Rune. Coffee table, Brasilia. Laminated plywood; h. 28cm (11in) w. 100–120cm (39$\frac{3}{8}$–47$\frac{1}{4}$in) l. 100–120cm (39$\frac{3}{8}$–47$\frac{1}{4}$in). Swedese, Sweden

Opposite: Benedini Associati. Washbasin, Woodline. Oak; h. 14cm (5$\frac{1}{2}$in) l. 70cm (27$\frac{1}{2}$in) d. 40cm (15$\frac{3}{4}$in) or h. 14cm (5$\frac{1}{2}$in) l. 100cm (39$\frac{3}{8}$in) d. 40cm (15$\frac{3}{4}$in). Agape srl, Italy

BEST SELECT

Hirokazu Sibata and Takeshi Kodera. SD-CX10 1-bit digital audio system. Plastics; h. 20.2cm (7^{7}8in) w. 33.2cm (13in) d. 16.8cm (6^{5}8in). Sharp Corporation, Japan

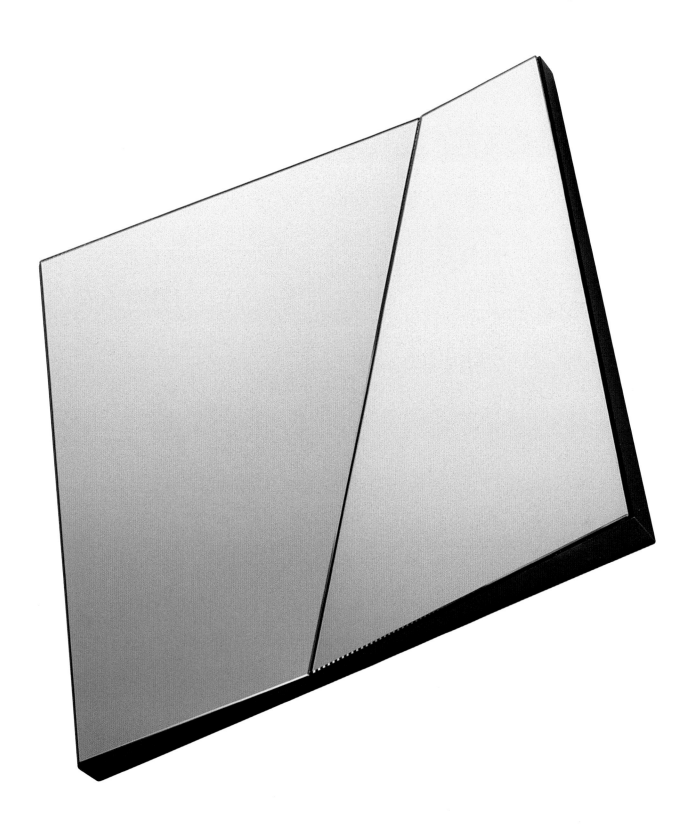

Giovanni Tommaso Garattoni. Modular mirrors. H. 40cm (15³₄in) w. 40cm (15³₄in). Tonelli srl, Italy

David Trubridge. Recliner, Sling. Oak, stainless steel; h. 60cm (23⁵⁄₈in) w. 60cm (23⁵⁄₈in) l. 237cm (93¹⁄₄in). Prototype

Christian Ghion. Lounger, Shadow. Thermoformed Corian®, chromeplated base; h. 70cm (27$\frac{1}{2}$in) w. 72cm (28$\frac{3}{8}$in) l. 180cm (70$\frac{7}{8}$in).
Cappellini SpA, Italy

Top: Kiyoshi Ohta. Mini disc player, Sony CMT-C7NT. Polycarbonate, stainless steel, aluminium die-cast, resin moulding; h. 125cm (49¹⁄4in) w. 145cm (57in). Sony Corporation, Japan

Above: Ferdi Giardini. Outside lamp, Teda. PMMA, zinc-plated metal; h. 90cm (35³⁄8in) diam. 5cm (2in). Oluce srl, Italy

Opposite: Norbert Wangen. Kitchen unit, K10. Stainless steel, light elements; h. 95cm (37³⁄8in) l. 366cm (144in) d. 72.5cm (28¹⁄2in). Norbert Wangen, Germany

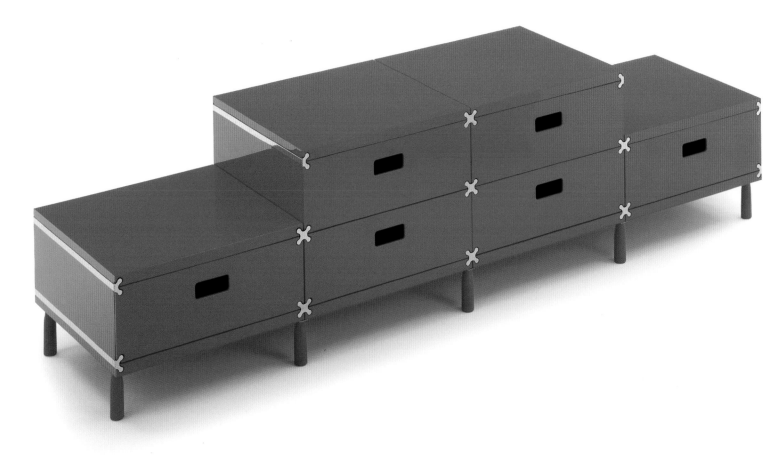

Opposite top: Werner Aisslinger. Modular drawer system, Plus Unit. Injection-moulded ABS, polished extruded aluminium; h. 21cm (8^{1}4in) w. 44cm (17^{3}8in) d. 44cm (17^{3}8in). Magis srl, Italy

Opposite centre and bottom: Werner Aisslinger. Modular container system, Case. Translucent acrylic, aluminium; h. 38–230cm (15–90^{1}2in) w. 60–80cm (23^{5}8–31^{1}2in) d. 39.4/65cm (15^{1}2/25^{1}2in). Interlübke, Germany

Below: Hannes Wettstein. Kitchen, Nomis. Dada SpA, Italy

2002 was the year of Eurocucina, and Rashid wanted to represent the trend for 'good taste' minimalist kitchens on offer by most of the major manufacturers. 'Trend' and 'taste' are words that do not appear in Karim's vocabulary, as the former brings to mind fashion, which is transitory, while the latter is objective. Yet it is hard to describe 'Nomis' in other terms. It was among the best quality on show – who would expect any less from Hannes Wettstein. However, it is hard to react to this example emotionally, which probably explains why Rashid did not have a lot to say about it. This kitchen demonstrates a simplified aesthetic with a minimum of expression. For the nuclear family, the kitchen used to be the most important room in the house, but today, in a time of rushed meals, convenience foods and the easy availability of good and cheap restaurants, its significance is being lost. Ever-smaller numbers of people eat together at home in social interaction, and the kitchen has lost its heart. A domestic kitchen should have the possibility of adapting and developing to everyday life, yet this exercise in restraint is a shrine to the immaculate.

Opposite above: Donato d'Urbino and Paolo Lomazzi. Hollow ware bowls, Double. Stainless steel; h. 5.8/7.3/9.5cm (2^{1}4/2^{7}8/3^{3}4in) diam. 20/25/32cm (7^{7}8/9^{7}8/12^{5}8in). Alessi SpA, Italy

Opposite below: David Mellor. Cutlery, Minimal. Stainless steel. David Mellor Design Ltd, UK

Above: Scott Henderson, Three-chambered serving dish, Twist. Chrome electroplated injection-moulded ABS; h. 4.5cm (1^{3}4in) w. 21.5cm (10^{1}2in) l. 44cm (17^{1}4in). Wovo and Smart Design Inc, USA

Opposite: Theo Williams. Telephone wire, OYO. ABS; diam. 8cm (3^{1}8in). Lexon, France

Above left: Paolo Ulian. Breadboard, Virgola. Polyethylene; h. 3cm (1^{1}8in) w. 30cm (11^{7}8in) l. 35cm (13^{3}4in). Zani&Zani, Italy

The breadboard is cut from one piece of polyethylene and has a large handle that rises from the base, making it easy to carry.

Above right: Apple Industrial Design Team. MP3 player, iPod. Polycarbonate, stamped stainless steel; h. 10.2cm (4in) w. 6.2cm (2.4in) d. 1.9cm (^{3}4in). Apple Computer Inc, USA

The iPod is the first MP3 player to hold 1,000 songs, five gigabytes of data, have an 8-hour battery and only weight 184 g (6^{1}2 oz).

Above: Hiroyuki Mastushima. Multimedia speaker, L'Oto. Polyester resin compound reinforced with glassfibre; h. 54cm (21$\frac{1}{4}$in) w. 48cm (18$\frac{7}{8}$in). Prototype; DMA, Japan

'L'Oto' sways and vibrates as the music plays, like flowers in the wind.

Opposite: Benjamin Hopf and Constantin Wortmann. Floor lamp, Lollipop. Aluminium; h. 90/150cm (35$\frac{3}{8}$/59in) w. 20cm (7$\frac{7}{8}$in) l. 20/100cm (7$\frac{7}{8}$/39$\frac{3}{8}$in). Prototype; Büro für Form, Germany

Above: Claudio Bellini. Desk, Scriba. Anodised aluminium, plexiglas; h. 87cm (34$\frac{1}{4}$in) w. 155/185cm (61/72$\frac{7}{8}$in) d. 90cm (35$\frac{3}{8}$in). YCAMI SpA, Italy

Opposite above: Alfredo Häberli. Seat with built-in desk, Solitaire. Wood, coldfoarm with flamefibre, fabric, chrome lacquer; h. 71cm (28in) w. 93cm (36$\frac{5}{8}$in) d. 62cm (24$\frac{3}{8}$in). OFFECCT, Sweden

Opposite below: Michele de Lucchi and Philippe Nigro. Table/tray, Volino. Moulded iron and leather sheets; h. 61cm (24in) w. 40cm (15$\frac{3}{4}$in) diam. 40cm (15$\frac{3}{4}$in). Poltrona Frau srl, Italy

Chris Ferebee. Bench, Joseph. Industrial wool felt; h. 40.6cm (16in) w. 45.7cm (18in) l. 60.9cm (24in). FiveTwentyOneDesign, USA

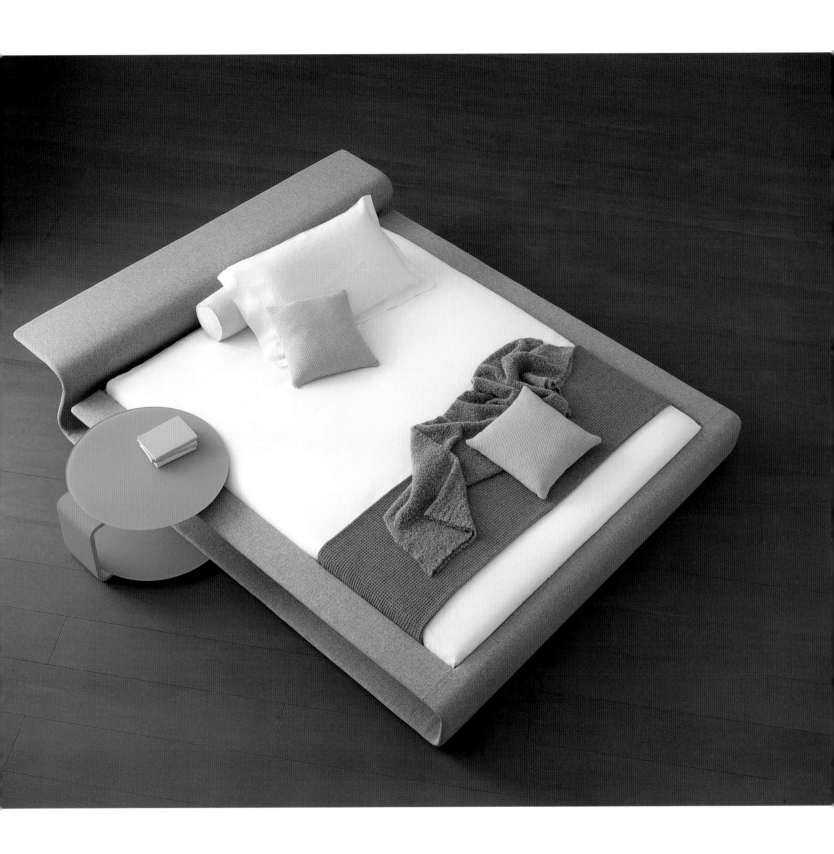

Patricia Urquiola. Bed, Clip. H. 77cm (30^{3}8in). Molteni SpA, Italy

Above: Cathrine Torhaug. Chaise longue, Flash. Fibreglass, stainless steel, leather; h. 83cm (32⁵⁄₈in) w. 146cm (57¹⁄₂in) d. 53cm (20⁷⁄₈in). Torh, Norway

Opposite: Biagio Cisotti and Sandra Laube. Bookcase system, Loop. Layered wood, chromed steel; h. 50cm (19⁵⁄₈in) l. 160cm (63in) d. 30cm (11³⁄₄in). BRF srl, Italy

Above: Christian Ghion. Armchair, Butterfly Kiss. Dacryl; h. 74cm (29$\frac{1}{8}$in) w. 60cm (23$\frac{5}{8}$in) l. 115cm (45$\frac{1}{4}$in). Prototype; Cappellini SpA, Italy

Oppposite: James Sung. Chaise longue, Glacier. Acrylic, aluminium; h. 76cm (29$\frac{7}{8}$in) w. 70cm (27$\frac{1}{2}$in) l. 180cm (70$\frac{7}{8}$in). Prototype; Sung Design Limited, UK

Anna von Schewen. Armchair, Hug. Solid bent beech, steel tubing; h. 74cm (28^{1}8in) w. 56cm (22in) l. 51cm (20in) d. 51cm (20in). Gärsnäs AB, Sweden

Julia Läufer and Marcus Keichel. Chair, Gap. Polypropylene; h. 78cm (30^{3}4in) w. 54cm (21^{1}4in) d. 53.5cm (21in). Prototype

Opposite top: Ricardo Bello Dias. Chaise longue, Memo. Chromium-plated metal, polyurethane foam; h. 65cm (25⁵⁄₈in)
w. 80cm (31¹⁄₂in) l. 155cm (61in). Nube, Italy

Opposite middle and bottom: Jeffrey Bernett. Armchair and sofa, Metropolitan. Nickeled or graphite metal, aluminium,
upholstery; armchair: h. 73cm (28³⁄₄in) w. 84cm (33¹⁄₈in) d. 83cm (32⁵⁄₈in), sofa: h. 70cm (27¹⁄₂in) w. 240cm (94¹⁄₂in) d. 97cm
(38¹⁄₄in). B&B Italia, Italy

Above: Pio e Tito Toso. Armchair, Slim. Acrylic, polyurethane, chrome-plated steel, leather; h. 66cm (26in) w. 65cm (85¹⁄₂in)
d. 76cm (29⁷⁄₈in). Frighetto Industrie srl, Italy

Below: C. Ballabio and A. Elli. Armchair, Waimea. Plywood, leather, chromed metal; h. 64cm (25$\frac{1}{8}$in) w. 80cm (31$\frac{1}{2}$in) l. 80cm (31$\frac{1}{2}$in). Emmemobili, Italy

Bottom: Peter Henriksen. Bed, Raft. Metal; h. 35cm (13$\frac{1}{2}$in) w. 163cm (64in) l. 205cm (80in). Innovation Randers A/S, Denmark

Opposite: Piergiorgio Cazzaniga. Chair, Fold. Lacquered wood, stainless steel; h. 78cm (30$\frac{3}{4}$in) w. 47cm (18$\frac{1}{2}$in) d. 50cm (19$\frac{5}{8}$in). Tagliabue srl, Italy

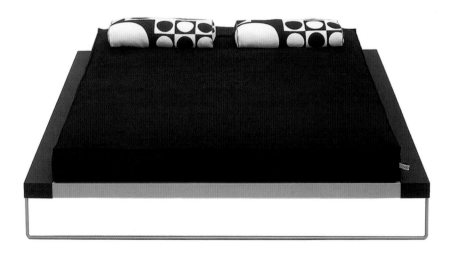

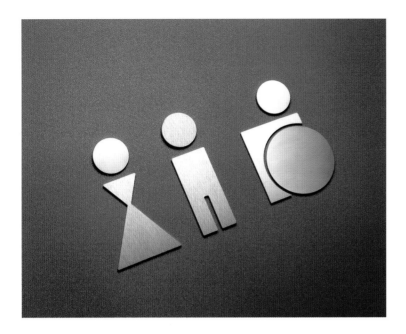

Opposite: De Vecchi Too range. Clockwise from top left: Ora Ïto. Salt and pepper cellar, Capsule. Dish, Module. Ronan and Erwan Bouroullec. Two-sided mirror, TV Mirror. Mirror. Wall ornament, Diana. Jean Marie Maussaud. Tray. Silver. De Vecchi, Italy

The nineteenth-century Paris Salon each year set its artists a theme to interpret, usually an historical subject or a scene from classic mythology. The results were exhibited in their hundreds, hung several high to line the walls of the Grand Palais. Because of the academic rigours and stipulations of the academy, there was little difference between the paintings displayed. Yet every now and then, to great public outcry, an adventurous individual would try to break the strict confines of tradition by experimenting with technique or subject matter – only to be summarily rejected. In 1863, Napoleon III authorized the Salon des Refusés, which is recognised as a landmark in the history of modern art as it gave a platform for artists who went on to form the Impressionist movement. Although not thematic, de Vecchi has similarly sought to liberate silver from the confines of traditional craftsmanship. By inviting designers such as Rodolfo Dordoni, Ora Ïto, Patricia Urquiola, Ludovica and Roberto Palomba, Ronan and Erwan Bouroullec and Jean Marie Maussaud more used to working with modern technology methods than a raw and semi-precious material, they have undressing silver of its formal appearance, making it more flexible, humane and contemporary. The collection is playful, imaginative and impressionistic.

Above: Andreas Winkler. WC symbols. Stainless steel; h. 11cm (4^{3}8in) w. 3cm (1^{1}8in). Phos Design, Germany

Charles O. Job. Double-sided bench, Vis-à-vis. Upholstered plywood, stainless steel; h. 80cm (31¹2in) w. 75cm (29¹2in) l. 225cm (88⁵8in). Prototype

Charles Job's ambition is to work with manufacturers who are concerned with giving form and expression to everyday, affordable objects. He was born in Nigeria but works in Switzerland, and believes that the strong Protestant leanings of the Swiss, combined with their equally strong artisan tradition, has fostered a high level of workmanship. A Quaker-like eye for the 'understated' has had a direct affect on his work over the past ten years. His 'Vis-à-vis' seat has a simplicity of form and a dematerialized elegance, which makes it interesting to see what his future designs will bring.

Above left: Theo Williams. Torch, Handy Light. ABS, aluminium; l. 14.8cm (5³4in) diam. 1.6cm (⁵8in). Lexon, France

Above right: Ian Clarke. Pendant light, Flute. Blown clear glass, polished aluminium, chrome wire; h. 39.5cm (15¹2in) diam. 11.5cm (4¹2in). Aktiva Systems Ltd, UK

As we exist more and more in a virtual world, our physical world is paradoxically developing a new importance. The furnishings around us are becoming special considerations: more beautiful, diverse and personal. Our future furnishings will be technological. Objects will become intelligent. Strangely there are still not enough examples of the merging of technology, furniture, architecture and objects, but eventually we will have a seamless marriage between these elements. High-tech and low-tech will merge. We are creating a new post-industrial digital language – a techno-aesthetic. I believe technology will combine with humans and our physical landscape. Designers are searching for new and 'smart' materials, novel production methods and ways of exploiting old and new machinery, and are using CAD/CAM software to develop digitally poetic proposals for our material landscape. Short-term production technology will change, making design less involved with the banalities of problem-solving, and more focused on emotional and pleasurable human experiences. Flexibility is the key to the future, where we can simultaneously be anywhere, and work, communicate and play in an omnipresent real-time existence. The design challenges will be two-fold, arising from changes in manufacturing and the democratization of technology. In the twenty-first century, digital 'desktop manufacturing' and variable non-serialized design will come to the fore. Where products can meet individualized criteria and taste, and as technology improves, objects and products and space will all be highly customizable, and possibly designed by any individual. Everything will be unique.

Techno

'An artist certainly cannot compete with a man on the moon in the living room.'
Kynaston McShine
'But a designer can.' Karim Rashid

Previous page: Karim Rashid. Chair, Alo. Polished aluminium plate, standard injection-moulded polypropylene, chromed steel; h. 82cm (32^{1}4in) w. 53cm (20^{7}8in) d. 55cm (21^{5}8in). Magis srl, Italy

Above: Janos Korban and Stefanie Flaubert. Chaise longue, Membrane. Electropolished spiral mould resistant welded stainless steel mesh; h.80cm (31^{1}2in) w. 60cm (23^{3}4in) l. 180cm (70^{7}8in). One-off

Opposite: Asymptote. Furniture system, Knoll A3, Urba. Steel; UV-cured tensile mesh screens, powder-painted wood, injection-moulded plastic; h. 175.5cm (69in) w. 175.5cm (69in) d. 273.3cm (108in). Knoll, USA

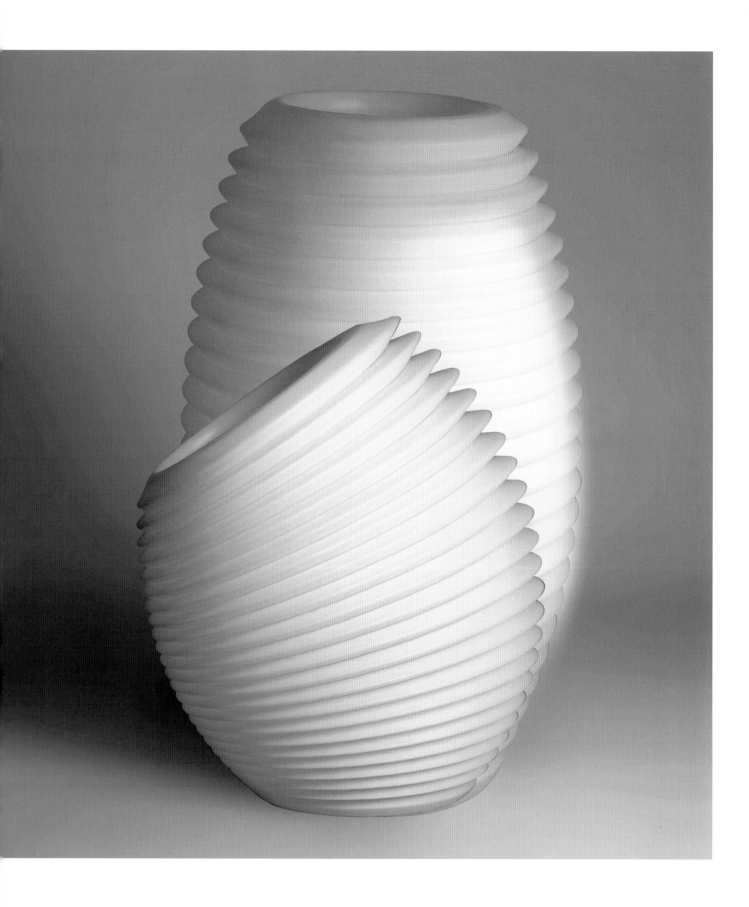

 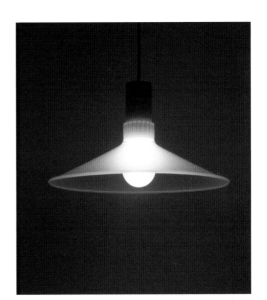

Opposite: Ron Arad. Expandable outdoor pot-planter, Toppot. Polyethylene; h. 108cm (42$\frac{1}{2}$in) diam. 73cm (28$\frac{3}{4}$in). Serralunga, Italy

Ron Arad has elevated the humble flowerpot using the technology of rotation moulding. The volume of the container is broken by the contracting bellow action of the zigzag section. A valve in the double-skin pot controls the amount of air trapped in the bellow and so freezes the height and inclination of the 'Toppot' at any given position. Alternatively, you can cut off the inner skin and let the earth that fills the pot act as a fixative for the volume and position.

Above: Sam Hecht. Pendant ceiling light, 12 Light Years. Polyurethane; h. 11cm (4$\frac{3}{8}$in) diam. 32.5cm (12$\frac{3}{4}$in). Limited batch production; Proverb, UK

'12 Light Years' is a lampshade made from a single polyurethane rubber moulding that slips over an ecologically sound bulb. The light will last for up to twelve years before needing to be changed.

Above: Ingo Maurer. Pendant lamp, Stardust. LED, glass. Ingo Maurer GmbH, Germany

Opposite: Francois Azambourg. Lamp, Yvette. PMMA optic fibre, metal; h. 150cm (59in) diam. 35cm (13³4in). Galerie Kreo, France

Azambourg uses fibre optics with lateral emission, which he braids together into a tube that can then be formed into a variety of shapes.

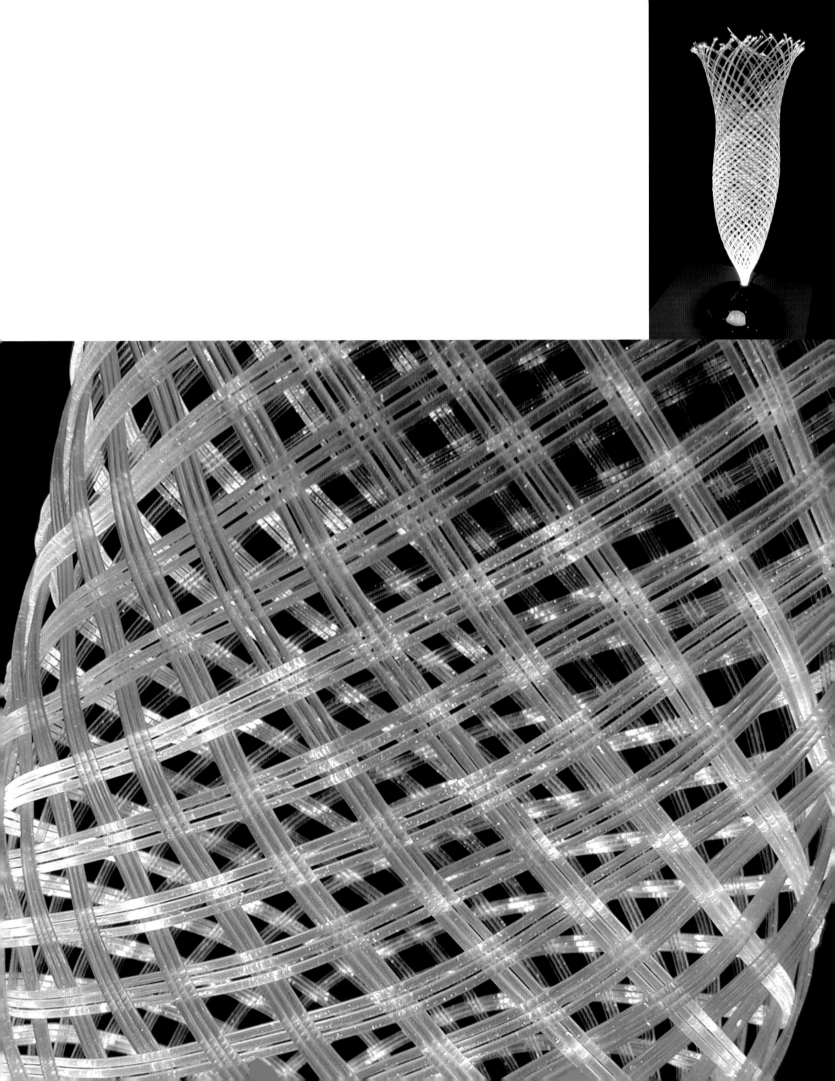

Tokujin Yoshioka. Chair, Honey-pop. Paper; h. 83cm (32⁵⁄8in) w. 80cm (31¹⁄2in) d. 83cm (32⁵⁄8in). Tokujin Yoshioka Design, Japan

Only a handful of objects impressed at last year's Milan Furniture Fair and one of the most lasting memories was at the Driade show on Via Montepoleone. Against a pristine white background, Tokujin Yoshioka had created a magical snowscape that actually crunched under foot. You had to pass by his 'Tokyo Pop' series (see page 87) before reaching the inner sanctum, where the lights were dimmed and spots outlined the ephemeral spectre of the Honey-Pop Chair. Tokujin states that 'One of the standards by which I assess myself is whether I can create a design that touches and amazes me. Design should express emotions such as surprise, joy and wonder. I strongly hope to create a design that I have never seen and which I want to see'. The construction of 'Honey-Pop' is reminiscent of the 1950s concertina Christmas decorations, but in reality it is a cellulose version of the exceptionally strong and light honeycomb material used in the aeronautical industry. This has been cut to a thickness of 1cm (³⁄8in) unfolded and is sat on to form an ergonomic shape which could only result from the pressure of hip on paper. It certainly does surprise (at the freshness of concept), fill you with joy (at the sheer poetry of its aesthetic) and wonder (at how can something that light and delicate take the weight of a fully-grown man). The chair reflects Yoshioka's interest in new materials, or using materials adaptively, lifting them beyond their conventional potential. The material suggests the form, which he then liberates from it: 'Material is not an end to itself. It is only after examining and studying it closely that I can envisage giving it concrete form. How it can be made even more interesting.'

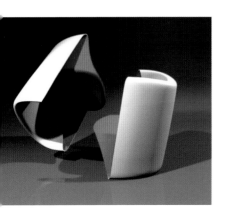  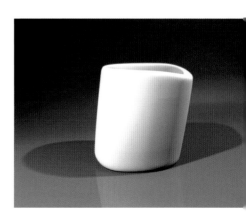

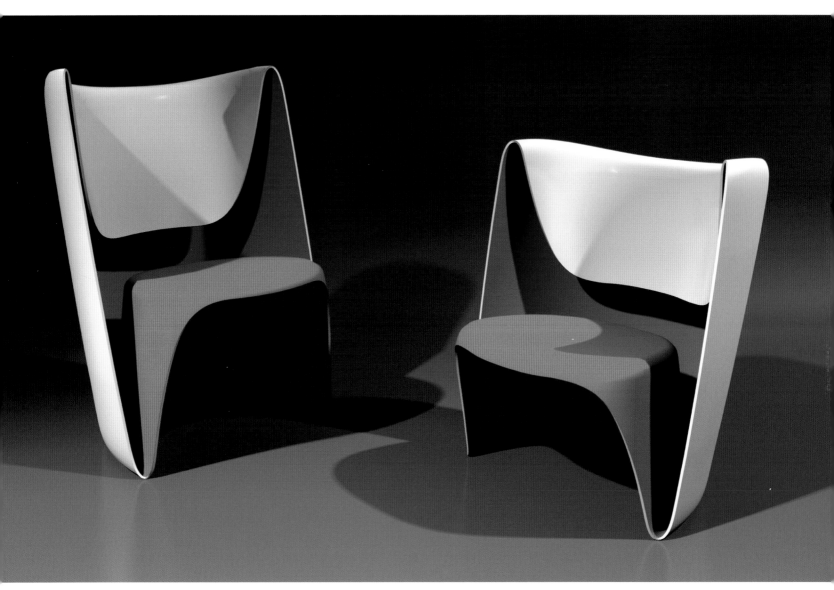

Ron Arad. Chair and armchair, Nino Rota e None Rota. Moulded polyethylene; chair: h. 42/88cm (16¹⁄2/34⁵⁄8in) w. 61 cm (24in)
d. 61cm (24in), armchair: h. 34/72cm (13³⁄8/28³⁄8in) w. 63cm (24⁵⁄8in) d. 61cm (24in). Cappellini SpA, Italy

Nathelie Jean. Bowls, La Ville Nouvelle. Stainless steel, silver, enamel. Design Gallery Milano, Italy

We are used to defining only sinuous, curved and soft shapes as 'organic'. In reality, many angular shapes can be found in nature. Nathelie Jean has combined high technology and craftsmanship to form a collection, 'La Ville Nouvelle', which explores the expressive possibilities of molecular nature through the observation of rotational geometric models such as DNA molecules, shells and crystalline structures. The process went from the manual folding of cardboard models, to the virtual, interpreting these shapes into CAD designs, and back to the manual, casting them (in this case the bowls) with the precision of a goldsmith in stainless steel and silver, and lined with coloured enamel.

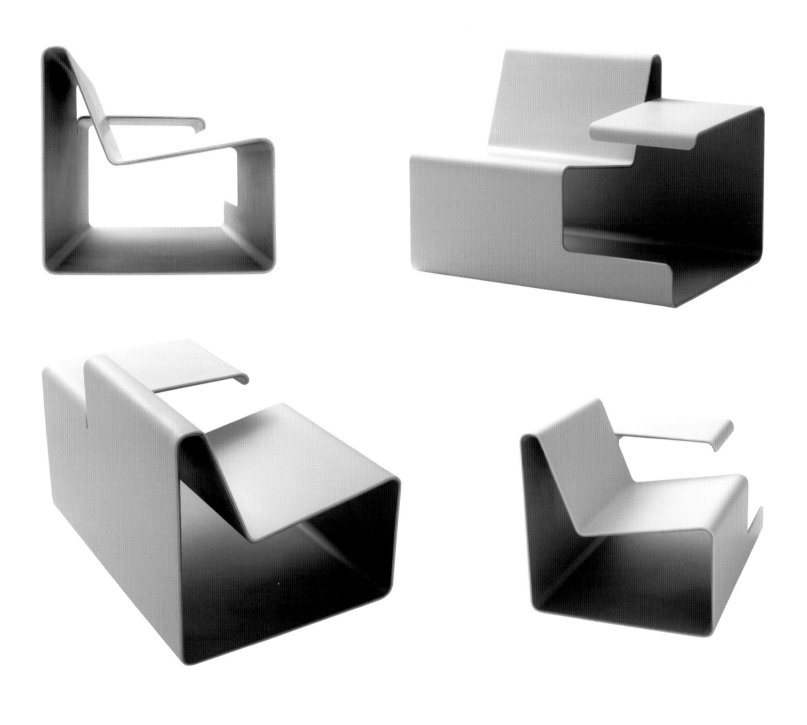

Catharina Lorenz and Steffen Kaz. Chair, Aspetto. Corian®, steel; h. 68cm (26^{3}4in) w. 88cm (34^{5}8in) d. 66cm (26in). Lorenz-Kaz, Italy

Proof that technology and innovative use of materials affects form, the complicated and sinuous shape of Catharina Lorenz and Steffan Kaz' 'Aspetto' was created on the computer and developed out of a single sheet of Corian®. The malleable quality of this material makes it suitable to interpret the new and complex designs being produced by CAD programmes.

Lievore-Altherr-Molina. Bench, Aero. Aluminium, stainless steel; h. 77cm (30^{3}8in) w. 65.7cm (25^{7}8in) l. 125cm (49^{1}4in) to unlimited d. 65.7cm (25^{7}8in). Sellex SA, Spain

Above: Tung Chiang. Chair, Audrey. Metal, moulded clear urethane rubber. Prototype

Below: Cathrine Torhaug. Bench, Lippy. Stainless steel, polyurethane; h. 50cm (19^{5}8in) w. 160cm (63in) d. 45cm (17^{3}4in). Limited batch production; Torh Møbler, Norway

Opposite: Karl-Otto Platz. Glass lighting system, Power Glass®. Cast resin laminated glass, integrated LEDs; h. 0.4–0.6cm (^{1}8–^{1}4in) w. 75cm (29^{1}2in) l. 250cm (98^{3}8in). Glas Platz, Germany

Ernesto Gismondi. Table lamp, Kaio. Painted metal, transparent polycarbonate ring diffuser; h. 61cm (24in) w. 12cm (4³⁄₄in) d. 52cm (20in). Artemide Design, Italy

Noriaki Takagi. Sony CD walkman, D-EJ1000. Magnesium alloy; h. 1.4cm (^{1}2in) w. 12.7cm (5in) l. 13.6cm (5^{3}8in). Sony Corporation, Japan

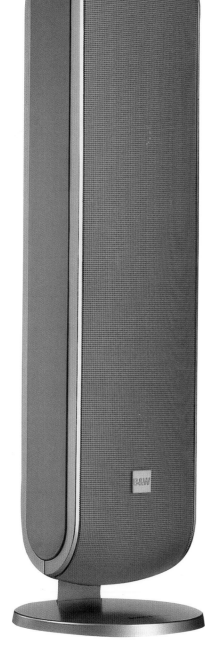

Above: Morten V. Warren. Loudspeaker, VM1. Moulded plastics, aluminium, steel; h. 52.2cm (20^{3}8in) w. 12.4cm (4^{7}8in) d. 9.3cm (3^{5}8in). B&W Loudspeakers Ltd, UK

Opposite: Ichiro Iwasaki. Remote control, mini component system Audio 405. ABS, aluminium, acrylic. Mutech, Korea

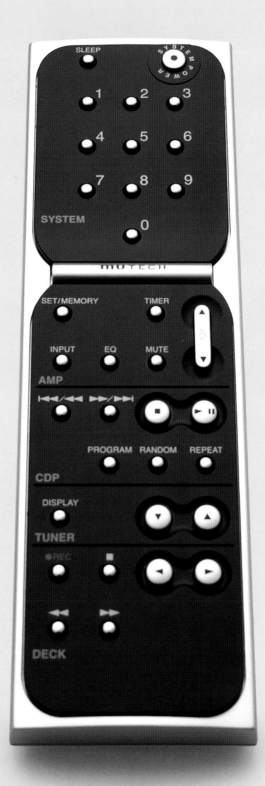

Opposite: Ernesto Gismondi. Table lamp, DuckLight. Zamak, diffuser, polycardonate coverbase. Artemide Design, Italy

'Duck Light' is Artemide's latest venture into e-lighting. The lamp is cold to the touch yet produces a warm light, which is switched on and dimmed by a mere brush of the fingers. It is highly adjustable and can be used to illuminate precision work.

Above: Apple Industrial Design Team. Computer, iMac G4. Polycarbonate, ABS, stainless steel; h. 50.9cm (20in) w. 41.5cm (16$\frac{3}{8}$in) d. 41.5cm (16$\frac{3}{8}$in) diam. 27cm (10$\frac{5}{8}$in). Apple Computer Inc, USA

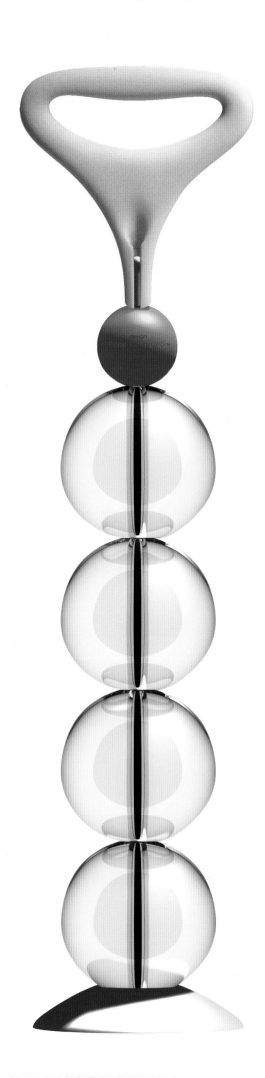

Opposite: Ross Lovegrove. Bowling set, Lawn Bowls. Magis srl, Italy

Magis have created a series of Post Computer Games designed on the computer. They wanted to remind us that design does not always have to be serious or clever, and that once in a while it's good to get back to basics, back to the games of childhood and tradition. Ross Lovegrove has designed a stand upon which four bowls and a jack ball can be strung and carried. The aim of the game is to roll your bowl as close as possible to the jack ball. This is harder than it sounds as the bowls are not perfectly spherical, and roll in curves.

Above: Ron Arad. Daybed/lounger, Oh Void 1. Nomex paper, Kevlar/carbon fibre; h. 123cm (48³⁄₈in) w. 47cm (18¹⁄₂in) l. 193cm (76in). Limited batch production; The Gallery Mourmans, The Netherlands

Below: Matthieu Manche. Garments, Fresh. Latex. Fresh, Japan

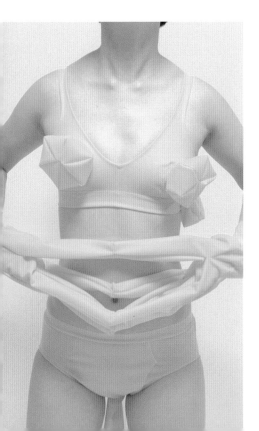
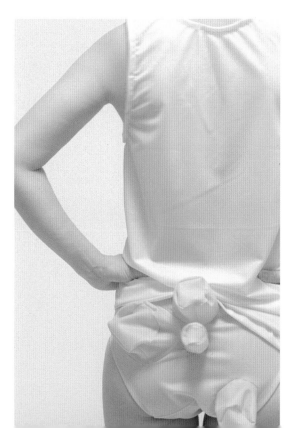
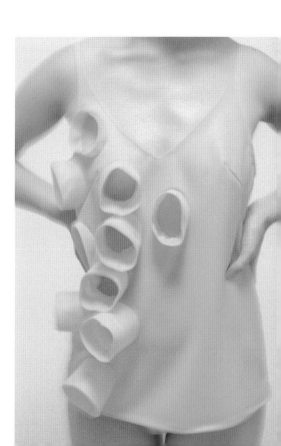

Below: Tom Dixon. Fruit bowls, Cesti. Polypropylene. Cappellini SpA, Italy

For the duration of the Milan Furniture Fair 2001, Tom Dixon and Domus collaborated to create an installation that involved both performance art and instant creativity. Set against the ornate backdrop of the deconsecrated church of San Paolo Converso, Dixon's plastic extrusion machine was visited by a host of designers including Michele de Lucchi, Michael Young, the Azumis, Ron Arad and Karim Rashid. All produced extraordinary feats of plastic engineering, from the sublime (Shin and Tomoko Azumi's over-sized dish sculpture) to the ridiculous (Young's urinating dog). Cappellini edited three of Dixon's designs, a chaise and chair called 'Spaghetti' and the 'Cesti' fruit bowls. He describes the process as contemporary knitting. The malleable material hardens almost immediately, the design being created instantaneously, each slightly different from the others. An added bonus is Dixon's discovery of a new technology he will not yet disclose; this changes the colour of the plastic in a moment. So if violet is in vogue today and acid green tomorrow, the products can be easily updated.

Opposite: Lauren Moriarty. Interior Cube. EVA rubber foam; h. 45cm (17^{3}4in) w.45cm (17^{3}4in) l. 45cm (17^{3}4in) d. 45cm (17^{3}4in). Limited batch production; Lauren Moriarty, UK

Lauren Moriarty is interested in working three-dimensionally, reinterpreting what we have come to expect from textile design. At once a fabric, a sculpture or a piece of furniture, 'Interior Cube' has been made from intricately cut and moulded EVA rubber foam.

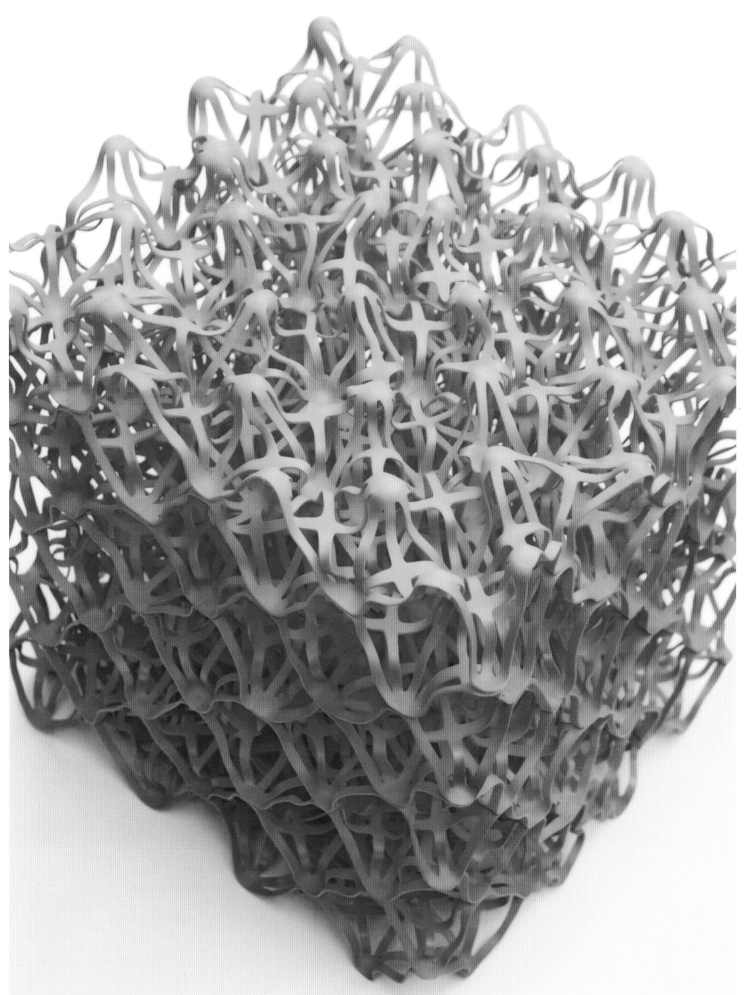

Lamberto Angelini. Trolley suitcase, Shuttle. Polypropylene; h. 80cm (32in) w. 60cm (24in) l. 35cm (14in) or h. 70cm (28in) w. 51cm (21in) l. 30cm (12in). Valigeria Roncato, Italy

Redman Instructor Suit and Student Suit. Vinyl-dipped foam. Macho Products Inc, USA

It looks like a cross between a 'Transformer' and the 'Indestructible Captain Scarlet', but there is nothing toy-like about the Redman Instructor Suit. The motto of Macho is 'Design with every advantage. If you limit your imagination, you limit your innovation. Remember, the people who use our products can be our best teachers.' The company was founded in 1980 by Grandmaster Soo Se Cho, who wanted to make his desire for greater safety in martial arts competitions a reality. Combining nearly thirty years of combat training and his artist's eye, he began designing Macho's original line of sparring gear. The idea behind the suit is that it should protect and cushion safety instructors or law enforcement officials, yet at the same time they should be able to feel some of the force of the blows thereby keeping the training as realistic as possible. The gear had to be light to aid free movement and not totally shock absorbent so that the trainee would not feel they were invincible. The manufacturing process used by Macho has advanced closed-cell foam production using direct injection-moulding technology, which gives these products exceptional detail, durability and reduced weight. Soft foam can now be used as an alternative to polyurethane/PVC. Rashid selected this example as he liked 'this idea of redefining and presenting a future world — a world where we can loose many traditions, rituals and myosis about how we should live, dress, think and work.'

Biographies

Eero Aarnio was born in 1932 in Helsinki, Finland. He studied at the Institute of Industrial Arts in Helsinki and began working with plastics in 1960, opening his own studio in 1962. Aarnio created two of the most famous chairs of the 1960s, the 'Globe' and the 'Gyro'. His designs are manufactured by Adelta. 36, 50, 51

Agnoletto-Rusconi is a partnership between Marzio Rusconi Clerici and Laura Agnoletto. Rusconi was born in Milan, Italy, in 1960 and trained at Milan Polytechnic. Agnoletto was also born in Milan, in 1963. After two years working in London, they returned to Milan in 1987. Clients include Memphis, Pallucco and FontanaArte. 66

Werner Aisslinger was born in Berlin in 1964. He founded his own company in 1993 and since then has carried out furniture and corporate architecture projects for companies such as Cappellini, Magis and Lufthansa. 182

Lamberto Angelini founded Angelini Design in Bologna in 1980. A graduate in mechanical engineering, he worked at the Volkswagen Style Centre in Germany, and from there moved into product design. The company now works for clients such as Acquaviva, Motobechane, BMW, Castelli, Mandarina Duck and Goldoni. 234

Richard J. Anuszkiewicz was born in 1930 in Erie, Pennsylvania, USA, and studied at the Cleveland Institute of Art, Yale University, and Kent State University. He was a leading figure in the Op art movement in the 1960s. 31

Ron Arad was born in Tel Aviv, Israel, in 1951. He studied at the Jerusalem Academy of Art, and at the Architectural Association, London. In 1981 he co-founded One Off. In 1988 Arad won the Tel Aviv Opera Foyer Interior Competition, and formed Ron Arad Associates in order to realize the project. Other projects include furniture design for Poltronova, Vitra, Moroso and Driade, and the interiors of the Belgo restaurants in London. In 2000 he had a major retrospective at the Victoria & Albert Museum, London. 212, 217, 231

Asymptote is an architectural design company founded in 1989 in New York by Hani Rashid and Lise Anne Couture. Rashid was born in 1958 in Cairo, Egypt, and Couture in 1959 in Montreal, Canada. Their projects range from urban design to computer generated environments and modular furniture systems. In 2000 they designed the Issey Miyake flagship store in New York. 211

François Azambourg was born in France in 1963 and studied industrial design in Paris. He has designed interiors for various Parisian shops, and lists Cappellini, Oluce and Mandarina Duck among his clients. 215

Shin and Tomoko Azumi studied industrial design at Kyoto City University of Art and the Royal College of Art, London. They founded Azumi's in 1995. In 1996 they were finalists in the Blueprint/100% Design Awards. 78

Enrico Azzimonti was born in 1966 in Busto Arsizio, Italy. He graduated in architecture from Milan Polytechnic in 1993, where he took a Master's in design in 1996. He then studied at the Domus Academy in 1998 and began working with Jordi Pigem. Jordi Pigem de Palol was born in 1968 in Banyoles, Spain. He trained in interior design at the University Menendez Pelayo in Barcelona, and at the Domus Academy in Milan. 102

Emmanuel Babled was born in France in 1967. He studied at the Istituto Europeo di Disegno in Milan, then worked for several years with Prospero Rasulo and Gianni Veneziano at Studio Oxido in Milan. His clients include Baccarat, Fine Factory, Wedgwood and Kundalini. 132, 133

Barber Osgerby is a design consultancy that was founded in 1996 by Edward Barber and Jay Osgerby. They have worked for companies such as Cappellini, Isokon Plus, Offecct and Asplund. Named best designers at I.C.F.F. in New York in 1998, they founded Universal in 2001 to oversee interior and architecture projects. 144

Claudio Bellini was born in Milan in 1963. He graduated in architecture and industrial design from Milan Polytechnic in 1990 and worked at Mario Bellini Associates until 1996. In 1997 he founded Atelier Bellini, a consultancy devoted to industrial design. His recent clients include Vitra, Artemide, Fiat, Venini and Driade. 190

Ricardo Bello Dias studied architecture and urbanism in Brazil. In 1994 he moved to London to attend the Architectural Association. He worked for architectural firms in Brazil, Milan and London, before opening his own studio in Milan in 1996. In 2002 he became art director of Vivendum in New York. 200

Juan Benavente Valero was born in Teruel, Spain, in 1973. After studying industrial design in Valencia he began working as a freelance designer in 1997, opening his own studio called Juanico Design. Benavente has designed furniture, lighting and products for firms including Sancal and Especta, and has worked on corporate identity. 61

Benedini Associati is a family firm founded by architectural designer Giampaolo Benedini, who was born in 1945 and studied at Milan Polytechnic. Benedini Associati designs products, and office, bathroom and kitchen furniture and accessories. Giampaolo is co-art director of Agape. 175

Markus Benesch was born in Munich in 1969. He studied in Birmingham in the UK, and later at the Domus Academy in Milan. Since then he has been active in the fields of interior and product design. He set up the 'Money for Milan' initiative in 1995 and since then has exhibited regularly. 68

Guglielmo Berchicci was born in Milan, Italy, and studied architecture at Milan Polytechnic. He has worked for the companies Kundalini, Giovanetti, Barneys New York and Vivienne Westwood, among others. 35

Jeffrey Bernett set up Studio B in New York in 1995. In 1996 he presented an award-winning range of furniture at the New York Furniture Fair. He works in many areas, including interior design, furniture, lighting and corporate image. His client list includes Authentics, B&B Italia, Cappellini, Dune and Ligne Roset. 200

Francesco Binfaré was born in 1939 in Milan. From 1969 he was a director at Cassina. In 1980 he set up the Centre for Design and Communication to produce works such as Wink by Toshiyuki Kita, Feltri by Gaetano Pesce and designs for De Padova by Vico Magistretti. His own client list includes Cassina and Edra. 160

Luca Bonato was born in 1962 in Marostica, Italy. He works from his family's workshop in Bassano del Grappa, and began specializing in designs in plexiglass in 1970. Over the last decade he has designed for Fusina. 104

Erwan and Ronan Bouroullec have collaborated since 1998. Erwan studied industrial design at the Ecole Nationale Superieure des Arts Appliques and the Ecole Nationale des Arts Decoratifs. Ronan graduated in applied and decorative art and has worked freelance since 1995, designing objects and furniture for Cappellini, Liaigre, Ligne Roset and Galerie Neotu. In 1999 he was awarded Best Designer at the ICFF in New York. 33

Boym Partners is a New York-based design firm. Constantin Boym was born in 1955 in Moscow, Russia, where he studied at the Moscow Architectural Institute. In 1985 he attended the Domus Academy in Milan. He set up his own company in 1986 in New York, and built up a client list that includes Alessi, Authentics and Swatch. Laurene Leon Boym was born in New York, where she studied at the School of Visual Arts and the Pratt Institute, graduating in 1993. Since 1994 she has been a partner in Boym Partners Inc. 54, 59

Büro für Form was founded in 1998 by the German designers Benjamin Hopf and Constantin Wortmann. They work in the fields of interior and product design, particularly lighting and furniture. Objects play with the perception of the viewer and aim to blend organic shape with geometric elements. In 2000 they exhibited for the first time in Milan and Cologne. 21, 189

Humberto Campana trained as a lawyer but in 1983 teamed up with his architect brother Fernando. Based in São Paolo, Brazil, the siblings were thrust into the limelight with their 1989 exhibition 'The Inconsolable'. The brothers produce political rather than merely functional objects. Their radical vision involves the use of traditional materials and industrial surplus to create an independent Brazilian design language. 111, 136–137

Mario Cananzi was born in Rimini, Italy, in 1958. He studied architecture at Florence University and industrial design at the Domus Academy in Milan. Cananzi has worked on architectural and furniture design projects for Edra and Disform, among others, as well as in the field of motorcycle design for firms such as Yamaha. 28

Chiara Cantono studied architecture and industrial design at Milan Polytechnic. She formed her own design studio in 1993, and her work is based on exploring with new materials and technologies. She has collaborated with companies such as Brunati Italia and the Busnelli Industrial Group. 19

Piergiorgio Cazzaniga was born in 1946. He studied at the IPSMA Institute in Milan, and Milan Polytechnic. In 1966 he was employed by Besana Mobili, and then joined Boffi Kitchens. He became an associate of Buttura, Massoni, Pelizza studio in 1971, which later changed its name to A&D. Clients include Dema and Liv'it. 203

Patrick Chia was born in Singapore and studied in Australia. He was discovered by Philippe Starck, who bought his first furniture design for the Mondrian Hotel. He is now based in Singapore. 147

Yuen Tung Chiang was born in 1966 in Hong Kong, where he studied graphic design and advertising. After a period working in advertising, Chiang took a course in environmental and product design at the Art Center College of Design in Pasadena, California. He is now based in Los Angeles. 222

Peter Christian was born in England in 1958 and studied at the Royal College of Art. He works in London with his wife Ruth Smart and designs lighting and furniture for a number of British firms. Recent exhibitions in Milan have drawn the attention of manufacturers such as FontanaArte, Elmar Flötotto and E&Y Tokyo. 70, 71

Cisotti-Laube is a collaboration between Biagio Cisotti and Sandra Laube that began in 1992. Biagio Cisotti studied architecture at the University of Florence. He has designed furniture and products for Alessi and BRF, among others. Cisotti co-founded the 'Made' exhibition and is art director for various companies. Sandra Laube studied at the Isia in Florence until 1993. She has designed products for firms such as BRF and Zeritalia. 195

Anna Citelli was born in Venezuela in 1964 of Italian parentage. In 1970 she moved to Italy, where she studied art in Milan and Bari. Since 1987 Citelli has lived in Milan, where she works freelance on a diverse range of projects, from Audi cars to pasta, and from L'Oreal cosmetics to Algida icecreams. 45, 62

Claesson Koivisto Rune is an architectural design company based in Stockholm and set up in 1995 by Mårten Claesson, Eero Koivisto and Ola Rune. Clients include Cappellini, David Design and Swedese. 37, 174

Ian Clarke was born in Yorkshire, UK, in 1973 and studied art and design technology at Leeds University. He worked for several years as a product designer. Clarke now designs products and lighting for Aktiva. 207

Claudio Colucci was born in Switzerland in 1965 and now lives and works in Paris and Tokyo. His design clients have included Martin Margiela, Issey Miyake, Habitat, Idée and Thomson Multimedia. 109

Mike Corbin was born in Gardener, Massachusetts. He started producing motorcycle accessories and parts in 1968. Drawing on his experiences as a Navy-trained electrician, he began developing electric mini-bikes in 1970. The first version of his 'Sparrow' electric car was manufactured in 1996. 97

Antonio Cos studied industrial design at the Raymond Loewy Institute in France, and graduated from the Advanced Institute for Artistic Industries in Florence, Italy, in 1999. Cos worked for Denis Santachiara and Fiorucci. He first exhibited at the Milan Furniture Fair as a freelance designer in 2002. 40, 46

Walter Craven studied sculpture at the Rhode Island School of Design, and trained in furniture design at the Center of Polish and European Art and Architecture. In 1996 he set up the studio Blank and Cables in San Francisco. He founded Walter Craven Design in 2002. 158

Lorenzo Damiani was born in 1972. He studied architecture at Milan Polytechnic and since then has taken part in many exhibitions and competitions. In 1999 he represented Milan in the design section at the Biennale of Young Artists and the Meditterranean. 146

DeCarloGuella is an architectural studio based in Milan, Italy, and established by Jacopo de Carlo and Andrea Guella. Their projects include furniture design and interior and architectural commissions for private clients and companies such as Napapijri and Futuro. 150

Degre Zero Architecture is a design firm set up in 1998 by David Serero, Elena Fernandez and Philippe Mohr. Serero studied at the School of Architecture Paris-Villemin and Columbia University, New York. Fernandez attended ETSAM and Columbia University. She worked with Peter Eisenman Architects. Mohr studied at the Bauhaus, Weimar, and Columbia University. He worked for Peter Eisenman and Arquitectonica. 94

Michele de Lucchi was born in Italy in 1951. He has won international renown as an architect, and was a leading force in the Cavart Group during the Radical Design years. De Lucchi was a founder member of the Memphis group. Clients include Arflex, Artemide, Kartell, Deutsche Bank, Olivetti and many others. 56, 191

Tom Dixon was born in Sfax, Tunisia, in 1959 and moved to the UK when he was four years old. He formed Creative Salvage with Nick Jones and Mark Brazier-Jones in 1985. His studio, Space, is where his prototypes and commissioned works – stage sets, furniture, sculpture and other objects – are made. His clients include Cappellini, Commes des Garçons and Terence Conran. He is head of design at Habitat UK. 232

Donato d'Urbino and Paolo Lomazzi were both born in Milan and began working together in 1966 on architecture, accessories and interior design. In the late 1960s they moved into furniture design. 184

Chris Ferebee was born in 1971 in Virginia Beach, USA, and has been based in New York since 1997, where he works in photography, mixed media assemblage, digital media, music and furniture design. In 1999 he formed FiveTwentyOneDesign with journalist and retailer Laurice Parkin. 192

Stanislav Fiala studied at CVUT in Prague and then at the AVU Architectural studio until 1986. Since the mid-1990s he has worked on architectural and interiors projects in Prague. He began collaborating with Daniela Polubedovová in 1997, who also studied at CVUT. They won the Mies van der Rohe prize in 2001. 55

Johannes Fuchs was born in 1967 in Bielefeld, Germany. He studied industrial design at the Hochschule für bidende Künste in Hamburg. In 1996 he joined Uwe Fischer in Frankfurt. He set up his own studio in 2000. 155

Giovanni Tommaso Garattoni was born in Rimini, Italy. In 1982 he co-founded the multimedia design studio Complotto Grafico, and in 1986 set up Il Bolidismo with other young architects. He has designed interiors for discos in Italy, and his clients include L'Oréal, Philips and Tonelli. 177

Gerhards & Glücker is a design partnership established by Carsten Gerhards and Andreas Glücker in 2000. Gerhards studied at the RWTH Aachen, the Bartlett School of Architecture in London and the Kunstakademie, Düsseldorf. He then worked for architectural firms in Germany and London. Glücker trained as an architect in Berlin and in urban studies at Bauhaus-Universitat, Weimar, and worked for various German architects. 152

Christian Ghion was born in Montmorency, France in 1958. He graduated from Etude de Creation de Mobilier, Paris in 1987. In 1998 he set up his own firm, concentrating on industrial and interior design for companies such as Neotu, 3 Suisses, Idée and Thierry Mugler. 124, 196

Ferdi Giardini was born in 1959 in Turin, Italy, where he now lives and works. He studied at the Accademia Albertina di Belle Arti di Torino, and his earliest works were in the field of theatrical set design. He has worked with the architects Derossi and Cucchiarati, and lists Oluce among his clients. 67, 180

Stefano Giovannoni was born in La Spezia, Italy in 1954 and graduated in architecture from the University of Florence in 1978. He founded King Kong Production to pursue research in design, fashion and architecture. Clients include Alessi and Cappellini. 41, 46, 49, 52, 101

Ernesto Gismondi was born in 1931 in San Remo, Italy. He studied at Milan Polytechnic and the Higher School of Engineering, Rome. In 1959 he co-founded Artemide SpA, of which he is president and managing director. In 1981 he was involved in establishing the Memphis group. 224, 228

Michael Graves studied architecture at the University of Cincinnati and Harvard University in the US. In 1960 he attended The American Academy in Rome. In 1964 he founded his own practice in Princeton, New Jersey. Graves has worked on projects for Disney and the Denver Central Library, and has received architectural design commissions around the world. He is currently Robert Schirmer Professor of Architecture, Emeritus. 138

Konstantin Grcic was born in 1965 in Germany. He trained as a cabinetmaker, then studied at the Royal College of Art, London, on a scholarship from Cassina. He worked for Jasper Morrison in 1990 and founded his own studio in Munich. Clients have included Agape, Cappellini, ClassicCon, Flos and Zeritalia. 158

Angelica Gustafsson studied product design at Beckmans School of Design in Stockholm, Sweden. She then undertook a post-graduate project with Orrefors-KostaBoda, and studied glass and ceramics design at the Royal College of Art in London. Gustafsson has a studio in Stockholm and works for clients such as Skruf. 116

Alfredo Häberli was born in Buenos Aires, Argentina in 1964, but moved to Switzerland in 1977. He studied industrial design in Zurich and worked as an installation designer at the Museum für Gestaltung until 1993. Since then he has worked freelance for companies such as Alias, Edra, Zanotta, Thonet and Driade. 191

Zaha Hadid is a London-based architectural designer whose work comprises urban, product, interior and furniture design. She was born in Baghdad in 1950 and studied at the American University in Beirut and the Architectural Association in London. She set up her own practice in 1979. In 1994 she received first prize in the competition to design the Cardiff Bay Opera House in Wales. 90

Dene Happell established the interior and product design firm Happell in 1998 in Glasgow, Scotland. The company has completed contract and domestic interior projects, as well as designing products for retail. 168

Sam Hecht was born in London in 1969. He studied at the Royal College of Art, after which he moved to Tel Aviv and joined the Studia group. His current collaboration with the IDEO group began in San Fransisco, but he also worked for them in Japan and in London, where he is head of industrial design. 213

Achim Heine was born in 1955 in Germany. After studying mathematics and design he took a degree in design at the Hochschule für Gestallung Offenbach. In 1985 he co-founded the design project 'Ginbande'. Heine is currently a partner of Heine Lenz Zizka Visual Communication, based in Frankfurt and Berlin, and head of Achim Heine Design in Berlin. His clients include Audi, Leica, Thonet and Vitra. 25

Scott Henderson was born in 1966 and lives in New York, where he is Director of Industrial Design for Smart Design. He has lectured on design internationally and his work is held in the permanent collections of the Cooper-Hewitt National Design Museum and the Chicago Athenaeum. 77, 185

Yoshiki Hishinuma was born in Sendai City, Japan in 1958. He studied at Bunka College of Fashion and then worked for Issey Miyake before going freelance. In the 1980s he became known for his uniquely shaped clothes, which often employed wind and air. In 1992 he produced his own line of clothing. 125, 127

Hive is a design partnership formed by Mark Dyson and Monika Piatkowski in 1998 in London. Piatkowski trained at The American University in London and at Glasgow School of Art. She worked as a graphic and interior designer and produces art works. Dyson studied architecture and environmental design at Nottingham University and Glasgow. Hive's clients include Habitat, Nicole Fahri and Harrods. 117

Kate Hume studied fashion design in London. She then became a stylist, designer and trend consultant in London and New York. Hume now works with film, corporate identity and furniture design companies. In 1999 she began producing glass objects. 92

Ichiro Iwasaki is a product designer born in Tokyo, Japan, in 1965. After working at the Sony Design Centre in Tokyo he moved to Italy. In 1995 he established Iwasaki Design Studio. Iwasaki began designing products for the Mutech brand in 1999, and has since become design director. 227

ixilab is a design company set up by Izumi Kohama and Xavier Moulin in 1999, and based in Shikoka Island and Tokyo, Japan. Kohama studied at Musashiro Art University in Tokyo. She worked in Milan and Paris for companies including Aldo Cibic and Fiorucci. Moulin trained at the University of Industrial Design in Paris. He co-founded Black Radish Urban Mobile Design Agency in 1995, and in 1996 he became art director for Aldo Cibic. ixilab conducts research into lifestyle and products, and distributes products through the Internet. 165

Nathelie Jean was born in 1963 in Montreal, Canada, where she studied architecture. Jean worked with the architect Peter Rose, as a television set designer, and collaborated on the Canadian 'Center for Architecture' for the Venice Biennale in 1992. She moved to Milan in 1988, where she worked with Sottsass Associates and Aldo Cibic. Jean opened her own studio in 1993. 218

Charles O. Job was born in Lagos, Nigeria, and studied architecture at Oxford Polytechnic in England. Job worked in various architectural practices, including the studio of Santiago Calatrava in Zurich, Switzerland. He runs his own furniture design studio, and lists Prototipo and FontanaArte among his clients. 206

Viktor Jondal was born in Stockholm, Sweden, and moved to New York to study fine arts at Parsons School of Design. Jondal works in the fields of industrial design and architecture. He has worked for Karim Rashid and as a consultant for companies such as Pepsi, Reebok and DaimlerChrysler. 148

Patrick Jouin worked Thomson Consumer Electronics from 1994 to 1999, and is now a freelance designer. He has designed interiors for the Plaza Athénee Hotel in Paris, and furniture for Ligne Roset, Fermob and xO. 32

Cordula Kafka was born in 1966 in Gladbeck, Germany. She studied sculpture and ceramic design at the Kunsthochschule in Berlin. Kafka opened her own studio in Berlin in 2001. 169

King-Miranda Associati was founded in 1976 by Perry King and Santiago Miranda. They work in product and interior design, and are active in industries from furniture and lighting to telecommunications. 26

Korban-Flaubert is a design partnership between the metal artist Janos Korban and Stefanie Flaubert, who trained as an artchitect. In 1990 Korban and Flaubert left Melborne in Australia, where they had been based, to travel, eventually settling in Stuttgart, Germany. Flaubert worked for architect Günter Behnisch, while Korban designed exhibition projects. In 1996 they set up a furniture design workshop in Sydney, Australia. 210

Marc Krusin studied furniture design at Leeds Metropolitan University. Following placements in the UK and Italy, he moved to Milan and collaborated with various clients, including Piero Lissoni's office. In 1998 he co-founded Codice 31. He works with Bosa, FontanaArte and Saporiti, among others. 156

Tsutomu Kurokawa was born in Aichi, Japan, in 1962. He studied design at Tokyo Design School in Nagoya and trained in the offices of Ics Inc and Super Potato Co Ltd. He co-founded H. Design Associates in 1992. They have worked on a series of boutiques in Japan. In 2000 Kurokawa established Out DeSign. 128

Julia Läufer was born in 1968 in Freiburg, Germany. She trained in industrial design, fashion and textile design in Berlin and London. Since 1993 she has worked for fashion houses and in the film industry. In 2000 she began working with Marcus Keichel and Stefan Gauss on furniture design. 199

Scot Laughton runs his own design studio in Toronto, Canada. Laugton has worked with companies such as Umbra, Nienkamper and Pure, and was a founding partner of the furniture manufacturing firm Portico. 106, 149

Arik Levy was born in Tel Aviv, Israel, and graduated from the Art Centre College of Design in Lausanne. He worked for a period in Japan before moving to Paris, where he set up his own studio, 'L'Design'. 157

Gabrielle Lewin and Hlynur Vagn Atlason form a New York-based design team. Atlason was born in Reykjavik, Iceland, in 1974. He has lived and worked in Copenhagen, Paris and New York, collaborating with companies such as IKEA and Totem Design. Lewin is a graduate of Parsons School of Design, New York. She was listed as one of the world's top twenty young designers by 'Wallpaper' magazine in 2001. 43

Lievore-Altherr-Molina is a design company founded in 1991 by Alberto Lievore, Jeanette Altherr and Manel Molina. Lievore studied architecture in Buenos Aires, then in 1972 founded the furniture firm Hipótesis. After moving to Barcelona in Spain he formed the Grupo Berenguer with other designers, with whom he also set up SIDI. In 1984 he established his own design studio. Altherr studied at Darmstadt and in Barcelona. Molina studied at the Barcelona EINA design school. Their clients include Andreu World, Disform and Thonet. 220

Piero Lissoni was born in 1956. He studied architecture at Milan Polytechnic and then worked for G14 studio, Molteni and Lema. In 1984 he co-founded a company, taking on product, graphic and architectural projects. Since 1986 he has worked with Boffi Cucine, Porro, Living Design and Matteograssi. He was appointed art director for Lema and then in 1995, for Cappellini. In 1998 he began working with Benetton. 153

K.C. Lo was born in Hong Kong in 1962 and moved to Britain in 1976. He studied electrical engineering at Middlesex Polytechnic, then specialized in industrial design at Central St Martins College of Art and Design in London. Since 1990 he has worked for firms including B&W Loudspeakers and Johnson & Johnson. 170

Lorenz-Kaz is a partnership between Catharina Lorenz and Steffan Kaz. Lorenz studied industrial design at the Fachhochschule Darmstadt. She then worked for companies including Via 4 Design and Sotsass Associati, before opening her own studio in 2000 in Milan. Her clients include FontanaArte, Siemens and Zeitraum. Kaz studied at the Hochschule der biblenden Kunste in Hamburg and at the Royal College of Art in London. He set up his own studio in Milan in 2001. They began working together the same year. 219

Ross Lovegrove was born in 1958 in Wales. He studied industrial design at Manchester Polytechnic, and later at the Royal College of Art, London. He worked for various design consultancies, including frogdesign in Germany. In 1984 he moved to Paris to work for Knoll International and became a member of the Atelier de Nimes. In 1986 he co-founded Lovegrove and Brown Design Studio, which was later replaced by Lovegrove Studio X. Clients include Louis Vuitton, Philips, Sony and Apple Computers. 94, 99, 103, 230

Matthieu Manche trained at the Ecole des Beaux Arts in Grenoble and the Institut des Hautes Etudes en Arts Plastiques in Paris. He has exhibited his textile designs in New York, Paris and Tokyo. 231

Mark Mann was born in Germany in 1968. He studied business administration, and now lives and works in Frankfurt. His freelance design work is focused on lighting objects. 82

Ilaria Marelli studied at Milan Polytechnic. She works in photography, furniture, communication and theatre design. She has collaborated with Cappellini, Bosa, Tucano and Opos. 76

Jean Marie Massaud was born in Toulouse, France, in 1966, and graduated from the Atelier-Ecole Nationale Superieure de Creation Industrielle in 1990. In 1994 he started his own studio concentrating on industrial and interior design for companies including Authentics, Baccarat, Lanvin, Magis and Yamaha. 20, 30

Hiroyuki Mastushima graduated in architecture from Nippon University in Toyko in 1992, then worked for Kazuhiro Ishii Architect and Associates and Kitayama & Company. In 1998 he co-founded DMA, where he is now director. They work in the fields of architecture, furniture, product and graphic design. 188

Ingo Maurer was born in 1932 on the island of Reichnau, Lake Constance, Germany; he trained in typography and graphic design. In 1960 he moved to the USA and worked as a freelance designer in San Francisco and New York, returning to Europe in 1963. Maurer founded Design M in Munich in 1966. 214

Ross McBride was born in 1962 in Pittsburgh, Pennsylvania, USA. He studied graphic design at the California Institute of the Arts. In 1985 he moved to Japan, where he continued his training. He opened his own studio in 1991. During the 1990s his work shifted towards product, furniture and interior design. 135

David Mellor trained as a silversmith. He built his reputation as a designer of cutlery, which is manufactured at his factory in Sheffield, UK. Mellor was appointed Royal Designer for Industry and his work can be seen in the Victoria & Albert Museum, London, the Worshipful Company of Goldsmiths and MoMA, New York. 184

Alessandro Mendini works in furniture, fashion, interior and architectural design. He is currently a consultant to Studio Alchimia. Mendini designed the Alessi House and the Groningen Museum in Holland. 47

Stefano Miceli was born in Hannover, Germany in 1968 and studied architecture at Milan Polytechnic. He worked for the furniture firm Formfuersorge in Hannover and produced his own wooden furniture collection in 1995. Since 1999 he has produced lighting and luminescent furniture pieces for Memphis and others. 129

Jason Miller was born in 1971 in New York. He studied at Indiana University, before taking a Master's at the New York Academy of Art in 1995. Miller worked for the artist Jeff Koons and then as art director for Ogilvy and Mather, before joining the studio of Karim Rashid. He opened Jason Miller Studio in New York in 2000. 73

Monkey Boys is a design studio founded in 1999 by Bertjan Pot and Daniel White. They work in the fields of furniture, packaging, textiles and lighting. In 2000 the company moved to Rotterdam. 63, 115

Carlo Moretti was born in Murano in 1934 and today lives in Venice. He attended law school in Padua, then in 1958 established the Carlo Moretti company with his brother Giovanni. They work in Murano crystal and their designs can be found in major museums in Europe and the US. 134

Mauro Mori was born in Cromona, Italy, in 1965, and lives in Parma. He trained in architecture but works mostly in carving, especially in wood. 105, 120

Lauren Moriarty was born in 1978 and studied multimedia textiles at Loughborough University, UK. In 2000 she won a placement at Reebok in Boston, USA, where she designed materials for footwear. Since then she has been commissioned by landscape architects Tim Lynch Associates and Aerian Studios. 233

MNO is a partnership between the industrial designer Jan Melis and sculptor Ben Oostrum. Melis studied at the Design Academy in Eindhoven, while Oostrum trained at the Arts Academy in Rotterdam. They founded their company in 1999 and their clients include Droog Design, Cor Unum and Totem Design. 91, 93, 150

Christiane Müller was born in 1965 in Hamburg, Germany. She studied at the Design Academy, Eindhoven until 1989, and worked as a trendspotter for the Nederlands Interieur Institut. Müller is one of the partners of the Müller en van Tol in Amsterdam. She designs textiles for various companies including Danskina. 110

Yoshitomo Nara was born in Aomori, Japan, in 1959, and is one of a generation of Japanese artists whose work shows the influence of popular culture such as Manga comics and television animations. 47

No Picnic is an industrial design house founded in 1993 in Stockholm, Sweden. Their products range from low-tech furniture to outer space projects for the International Space Station. 98

Fabio Novembre was born in Italy in 1966. He trained as an architect and now lives and works in Milan. He is best known for his imaginative interiors for bars, shops and restaurants. He is also active in the field of product design and has collaborated with Cappellini. 16

Carl Öjerstam studied industrial design at the University College of Arts, Crafts and Design in Stockholm. He established his own studio, and designs lamps, kitchenware and other products for IKEA. 105, 171

Elizabeth Paige Smith studied at the School of Fine Arts at Kansas University. She then worked for Hirsch Bedner and James Northcutt Associates, who specialize in hotel design. She first showed her furniture designs at the New York International Furniture Fair in 1996. 69

Adrian Peach studied industrial design at Manchester Polytechnic, then worked as a furniture designer for Antonio Citterio. Peach set up his own studio in Milan in 1996. He is a consultant for Future Concept Laboratory and the Domus Academy Research Centre in Milan. Clients include Porro, Felicerossi and Segno. 83

Terri Pecora was born in 1958 in Idaho, USA, and attended the Art Center College of Design in Pasadena, California. In 1989 she moved to Italy to study product design at the Domus Academy in Milan. She set up her own studio in 1991, and has designed for DNA, BRF, Edra and Zanotta. 81

Paolo Pedrizzetti was born in 1947 and studied architecture at Milan Polytechnic. In 1978 he began working as a product designer with Davide Mercatali, and in 1982 they opened a studio. He founded Paolo Pedrizzetti & Associati with his wife Raffaella Mattia in 1988, producing tableware, lighting and bathroom accessories. 57

Gaetano Pesce studied at the University of Venice. In 1959 he founded the art research movement 'N Group' in Padua, and in 1971 set up La Compagnia Bracciodiferro to manufacture experimental objects. He moved to New York in 1983, where he branched out into architecture and founded Fish Design and Open Sky. 60

Olivier Peyricot studied at the Ecole Supérieure de Design Industriel. Since 1995 he has taken on diverse projects and collaborated with experimental studios such as ixilab in Japan, as well as with firms such as Edra. 95

PLH Design was established in Denmark in 1985 by the architect Steen Enrico Andersen. PLH designs products for clients worldwide, including Panvision, and has undertaken architectural commissions. 89

Geralt Ploegstra began working as a freelance designer in Rotterdam, The Netherlands, in 2000, under the company name of Univorm. Ploegstra researches and designs movable objects and furniture. 166

David Quan was born in Vancouver, Canada. Following his studies at Humber College, he worked for The Axis Group on sustainable design strategies, corporate identities and play-learning based design models. He then designed modular office furniture and co-founded Fusion Interactive. Since 1996 he has worked for Umbra. 142

Radi Designers was founded in Paris by Florence Doléac Stadler, Laurent Massaloux, Olivier Sidet, Robert Stadler and Claudio Colucci. Florence Stadler was a design consultant for Sommer Allibert until 1996. Massaloux and Sidet both worked for Philippe Starck in 1996 and for Thomson Multimedia. Robert Stadler was born in Austria and taught at the Academy of Applied Arts in Vienna from 1994 to 1997. 84

Nick Rennie was born in 1974 in Papua New Guinea, but grew up in Australia. He studied industrial design at Royal Melbourne Institute of Technology and was involved in the Melbourne Movement, which brought together local young designers. Rennie established his own furniture company called Happy Finish Design. 118

Von Robinson studied at Parsons School of Design in New York. He worked with Eva Zeisel on interiors and products, before founding VRiD with his wife Miho Suzuki. Clients include Frighetto, Moroso and Idée. 107

Andrea Ruggiero was born in 1969 in Italy. In 1987 he moved to New York to study at Parsons School of Design, and then attended the Domus Academy in Milan. From 1995 to 1997 he worked as an assistant to Constantin Boym, designing for Vitra and Authentics. In 2001 he established Studio Plus+. 78

Thomas Sandell was born in 1959 in Finland. He studied architecture in Stockholm and in 1989 set up his own studio. He has worked on interior and product design for clients such as Cappellini, IKEA and Swiss Air. 34

Denis Santachiara was born in Capagnola, Italy, in 1950. He began designing for a car company while also making conceptual art. In 1990 he founded Domodinamica with Cesare Castelli, producing experimental design. He has designed lights and furniture for FontanaArte and Campeggi, among others. 79, 161

Teresa Sapey studied architecture in Italy, and then fine art at Parsons School of Design. In 1990 she moved to Madrid, where she has worked for such companies as Absolute Vodka, Hugo Boss and Walt Disney. 126

Juergen R. Schmidt was born in 1956. He established Design Tech in 1983 and is author of several books. He has won the Braun Award for technical design and the VDID award of design for the disabled. 121

Inga Sempé studied at Les Ateliers-ENSCI in Paris and took a fellowship at the Villa Medici, Academy of France in Rome in 2000. Her designs have been manufactured by Cappellini, Baccarat and Via. 151

Jerszy Seymour studied engineering design at South Bank Polytechnic in London, then industrial design at the Royal College of Art. Now based in Milan, he has worked for companies such as Magis, Idée and Smeg. 63

Matt Sindall studied at Kingston Polytechnic. In the early 1980s he worked as an interior and set designer. He began designing furniture, packaging and accessories, and in 1988 co-founded O2, which produces environmentally friendly design. His clients range from Renault to 3 Suisses. 134

Paul Smith opened his a boutique in 1970 at age 23 in his hometown of Nottingham, UK, stocking his own designs alongside those of international designers. He launched his menswear label in 1976, followed by a childrens' clothing line in 1991 and a womenswear collection 1994. Paul Smith now has over 250 shops worldwide. 130, 131

Kivi Sotamaa studied spatial and furniture design at the University of Art and Design (UIAH) in Helsinki, Finland, and architecture at the Helsinki University of Technology (HUT). He is a partner of Ocean North, which investigates experimental architecture, urban and product design. 88

Rainer Spehl studied furniture design at the Royal College of Art in London. He designed office and domestic furniture for Dealerward, exhibitions for clients such as Ericsson, and has developed products for Iceberg and the fashion label RUDE. His own product range was launched in 2002. 58

Philippe Starck was born in Paris in 1949 and trained at the Ecole Camondo in Paris. He has been responsible for interior design schemes for Francois Mitterand's apartment and multi-purpose buildings such as the offices of Asahi Beer in Tokyo. As a product designer he collaborates with Alessi, Baum, Driade, Flos, Kartell and Vuitton. From 1993 to 1996 he was worldwide artistic director for the Thomson Consumer Electronics Group. In 1999/2000 he finished two central London hotels for the Ian Schrager group. 22, 90, 96, 102

Christopher Streng studied design and interior architecture at the Milwaukee Institute of Art and Design. In 1996 he began working as a freelance designer in Milan, Italy and the US, and in 1997 he launched a company to manufacture ceramics, resins and plastic products. He set up Christopher Streng Inc in 1999. 23, 108

James Sung worked for computer design companies in Taipei before moving to London to study industrial design at Central St Martin's School of Art and Design. In 2000 he established Sung Design Limited. 197

Terra Design was established in 1993 in São Paulo, Brazil, by Lars Jorge Diederichsen and Fabiola Duva Bergamo. Diederichsen worked as a joiner in Germany before studying at FH Kiel. He joined Raul Barbieri, then worked for Mexican companies, before co-founding Terra Design. Bergamo studied at the University Mackenzie, São Paulo, attended the Domus Academy, and worked for designers in Italy and Germany. 145

Renaud Thiry trained in design and applied arts at De Montford University in Leicester in the UK, then studied industrial design at ENSI/Les Ateliers. Since 1995 he has worked for companies such as Ligne Roset and Habitat, as well as producing his own designs under the name 'flandesign'. 74

Cathrine Torhaug was born in 1967 in Norway. She trained at the Glasgow School of Art from 1989 to 1993, and has worked as an interior and furniture designer. She established Torh Møbler in Oslo. 194, 222

Pio e Tito Toso are designers and architects who have collaborated since 1996. Their furniture, installation and graphic design clients include Artemide, Frighetto, Magis, Foscarini and Venini. 201

Fabiano Trabucchi graduated from Milan Polytechnic in 1995. His work is focused on interior design and illumination. He also designs furniture for a number of companies including Pierantonio and Bonacina. 29

David Trubridge studied naval architecture at Newcastle University, then self-trained in woodwork, taking commissions from the Victoria & Albert Museum, amongst others. From 1982 to 1986 he and his family sailed a yacht to New Zealand, building furniture in the Caribbean and Tahiti en route. Since then he has exhibited across the world and his work is manufactured by Cappellini. 179

Paolo Ulian was born in 1961 and studied painting in Carrara followed by industrial design in Florence. From 1990 to 1992 he worked at Enzo Mari's studio, then founded Paolo Ulian Industrial Design. He collaborates with firms such as Driade, Progetto and Zani&Zani. 75, 187

Patricia Urquiola studied architecture in Madrid and then at Milan Polytechnic. Since 1990 she has worked with Vico Magistretti, Marta de Renzio and Emanuela Ramerino. In 1998 she joined Lissoni, working with Cappellini and Cassina. Recent clients include Moroso, Fasem, Bosa and Tronconi. 100, 143, 193

Meike van Schjindel was born in Utrecht, the Netherlands, in 1973. She studied at Chabot College, California, at UC Extension, San Francisco, and at Hogeschool voor de Kunsten, Utrecht, where she now has her studio. 48

Guido Venturini was born in 1957 in Italy. He trained as an architect in Florence, where he still teaches interior decoration. He has worked with the designer Stefano Giovannoni, with whom he founded King Kong Production. Venturini is also co-founder of Bolidism. His clients include Alessi and Bianchi & Bianchi. 132

Anna von Schewen studied at Gothenburg University of Design and Crafts and the University of Art and Design in Helsinki. In 1995 she graduated from the University College of Art, Craft and Design in Stockholm. Von Schewen worked for Pelikan Design in Copenhagen until 1997, when she went freelance. 198

WAAC's is a Dutch design consultancy with specialists in industrial and interior design. Its partners are: product designer Joost Alferink, born in 1964 and graduate of the Technical University in Delft; interior architect Joost van Alfen, born 1962, who trained at Hogeshool Rotterdam en Omstreken; design engineer Marcel Jensen, born 1968; product designer Lisa Smith, born in 1974; and Roy Gibing, born 1970. 44

Marcel Wanders was born in 1963. He studied at various colleges of design in the Netherlands and in 1995 opened his own studio, Wanders' Wonders. Clients have included Cappellini, Droog Design and British Airways. In 2000 Wanders set up moooi, a production company that embraces the unconventional. 42, 159

Nobert Wangen was born in 1962 in Prum, Germany, and served a carpenter's apprenticeship before studying sculpture and architecture in Dusseldorf, Aachen and Munich. In 1991 he graduated in architecture from the Technical University in Munich and worked as a set designer. In 1995 his folding armchair Atilla was selected for the show 'Die Neue Sammlung Munchen'; it was then acquired for the Vitra Design Museum. 181

Morten V. Warren studied furniture and product design at Kingston University, then worked for Philippe Starck and Aldo Cibic. He founded the design firm Native in London in 1995. 226

Hannes Wettstein was born in Switzerland in 1958 and started out as a freelance designer. He then joined the Eclat Agency as a partner. In 1993 he co-founded 9D Design. Notable projects include the Swiss Embassy in Tehran and Grand Hyatt Hotel in Berlin. His clients include Cassina, Shimano and Artemide. 183

Theo Williams studied in Bristol and Manchester, graduating in industrial design in 1990. He has developed products for Prada, Armani, Calvin Klein and Tronconi, and is Art Director for Lexon, France. 186, 207

Andreas Winkler studied architecture in Braunschweig/Perugia and Vienna. In 1986 he opened his own studio in Karlsruhe, developing products for FSB, Tecnolumen, Vitra and others. He set up Phos Design in 1993. 78, 205

Jane Worthington studied at Cumbria College for Art and Design and De Montford University in Leicester. She then worked for Philips Corporate Design and established her own studio in 1995. 17

Sean Yoo worked as a city planner before training in industrial design at the Art Center College of Design in Pasadena, California. In 2000 he co-founded Apt 5 Design in Matera, Italy. 154

Tokujin Yoshioka studied at the Kuwasawa Design School and then worked for Issey Miyake. Since 1992 he has operated freelance and lists BMW and Shiseido among his clients. In 2000 he established Tokujin Yoshioka Design. 89, 216

Suppliers

Adelta, Friedrich-Ebertstrasse 96, D-46535 Dinslaken, Germany. T. +49 (0)2064 40797 F. +49 (0)3064 40798 E. adelta@t-online.de W. www.adelta.de

Agape srl, Via Po Barna 69/70, 46031 Correggio, Italy. T. +39 0376 250311 F. +39 (0)376 250330 E. info@agapedesign.it W. www.agapedesign.it

Agnoletto-Rusconi Architettura e Design, Via Amedei 3, Milan 20123, Italy. T. +39 (0)2 862 623 F. +39 (0)2 862 623

Aktiva Systems Ltd, 10B Spring Place, London NW5 3BH, UK. T. +44 (0) 20 7428 9325 E. info@aktiva.co.uk

Alessi SpA, Via Privata Alessi 6, 28882 Crusinallo (VB), Italy. T. +39 (0)323 868611 F. +39 (0)323 641605 E. info@alessi.com W. www.alessi.com

Apple Computer, Inc, 1 Infinite Loop, Cupertino 95014, CA, USA. T. +1 408 974 9202 W. www.apple.com

Apt 5 Design, Via Colgangiuli 5, 75100 Matera, Italy. T. +39 339 4573999 F. +39 333 8261999 E. apt5@apt5design.com W. www.apt5design.com

Artekno-Eps OY, Aakkulantie 46, FIN-36220 Kangasala, Finland. T. +358 (0)3 244 7638 F. +358 (0)3 244 7602 W. www.artekno.fi

Artemide SpA, Via Canova 34, Milan 20145, Italy. T. +39 (0)234 96 11 F. +39 (0)234 53 82 11 E. pr@artemide.com W. www.artemide.com

Asymptote, 561 Broadway 5A, New York NY 10012, USA. T. +1 212 343 7333 F. +1 212 343 7099 W. www.asymptote.net

Shin and Tomoko Azumi, 953 Finchley Road, London NW11 7PE, UK. T. +44 (0)20 8731 9057 F. +44 (0)20 8731 7496 E. mail@azumi.co.uk

Enrico Azzimonti, c.so Semmione, 100 21052 Busto Arsizio, Italy. T. +39 (0)331380673 F. +39 (0)331380673 E. e.azz@working.it

Emmanuel Babled, Via Segantini 71, 20 143 Milan, Italy. T. +39 (0)2 58 11 11 19 W. www.babled.net

Baccarat, 30 bis rue de Paradis, 75010 Paris, France. T. +33 (0)1 40 22 11 50 W. www.baccarat.fr

B&B Italia, Strada Provinciale 32, 22060 Novedrate (CO), Italy. T. +39 (0)31 795 111 W. www.bebitalia.it

Bart Design, Via A. Fratti 19, 56125 Pisa, Italy. T. +39 (0)50 220 12 33 F. +39 (0)50 220 64 84 E. info@bartdesign.it W. www.bartdesign.it

Bd Ediciones de Diseño, Mallorca 291, 08037 Barcelona, Spain. T. +34 93 458 69 09 F. +34 93 207 36 97 E. export@bdbarcelona.com W. www.bdbarcelona.com

Benesch Project Design, Mariannenplatz 1, 80538 München, Germany. T. +49 89 228 52 62 F. +49 89 228 57 14 E. info@moneyformilan.com W. www.moneyformilan.com

Blå Station AB, Box 100, S-296 22 Åhus, Sweden. T. +46 (0)44 24 90 70 F. +46 (0)44 24 12 14 E. info@blastation.se W. www.blastation.se

Bonacina, Via S. Andrea 20A, I-22040 Lurago D'Erba (CO), Italy. T. +39 (0)31 699225 W. www.bonacinapierantonio.it

Boym Partners Inc, 55-59 Chrystie Street #14, New York, NY 10002, USA. T. +1 917 237 0880 E. cboym@boym.com W. www.boym.com

Bozart, 320 Race Street, Philadelphia, PA 19106, USA. T. +1 215 627 2223 F. +1 215 627 2218 E. staff@bozart.com W. www.bozart.com

BRF srl, Loc. S. Maziale 21, I-53034 Colle Val d'Elsa (SI), Italy. T. +39 (0)577 929418 E. biancucci@brfcolors.com W. www.brfcolors.com

Brunati Italia, Via Catalani 5, Lissone 20035 (MI), Italy. T. +39 24 563 31 F. +39 24 562 67 E. infobrunati@cinova.it W. www.brunatiitalia.it

Büro Für Form, Hans Sachs Strasse 12, D-80469 München, Germany. T. +49 (0)89 26 949 000 E. info@buerofuerform.de W. www.buerofuerform.de

Burton Snowboards, 80 Industrial Parkway, Burlington 05401, VT, USA. T. +1 802 651 0462 F. +1 802 660 3250 E. info@burton.com W. www.burton.com

B&W Loudspeakers Ltd, Dale Road, Worthing BN11 2BH, West Sussex, UK. T. +44 (0)1903 221557 F. +44 (0)1903 221550 W. www.bwspeakers.com

Canon Inc Design Centre, 3-30-2 Shimomaruko, Ohta-Ku, 146-8501 Tokyo, Japan. T. +81 3 3758 2111 F. +81 3 5482 9852 W. www.canon.co.jp

Cappellini SpA, Via Marconi 35, 22060 Arosio (CO), Italy. T. +39 (0)31759111 F. +39 (0)31763322 E. cappellini@cappellini.it W. www.cappellini.it

Cassina SpA, Via L. Busnelli 1, I-20036 Meda/Milan, Italy. T. +39 (0)362372 1 F. +39 (0)362342246. E. info@cassina.it W. www.cassina.it

Tung Chiang, 818 Tularosa Drive #7, Los Angeles, CA 90026, USA. T. +1 626 390 8967 E. tungdont@hotmail.com

Christian Stuart Partnership, 51 Arthur Court, Charlotte Despard Avenue, London SW11 5JA, UK. T. +44 (0)20 7622 8096 F. +44 (0)20 7627 0744 E. idy@peterchristian.com W. www.peterchristian.com

Anna Citelli, Via Fumagalli 1, 20143 Milan, Italy. T. +39 (0)2 83241521 E. anncitel@tin.it

ClassiCon GmbH, Perchtinger Strasse 8, D-81379 München, Germany. T. +49 89 7481330 F. +49 89 7809996 E. rfvp@classicon.com W. www.classicon.com

Cor Unum, Gruttostraat 9-11, 5212 VM s-Hertogenbosch, The Netherlands. T. +31 (0)73 691 14 99 F. +31 (0)73 614 27 06 E. info@corunum.com W. www.corunum.com

Corbin Motors Inc, 2350 Technology Parkway, Hollister, CA 95023, USA. T. +1 831 635 1033 F. +1 831 635 1039 E. mda@corbinmotors.com W. www.corbinmotors.com

Antonio Cos, Viale Corsica 57/A, 20133 Milan, Italy. T. +39 (0)2733078 F. +39 (0)2733078 E. sophietonio@hotmail.com

Covo srl, Via degli Olmetti 3/b, 00060 Formello (Rome), Italy. T. +39 (0)6 90400311 F. +39 (0)6 90409175 E. mail@covo.it W. www.covo.it

Walter Craven, PO Box 410537, San Francisco, California 94141, USA. T. +1 415 648 3872 F. +1 415 648 4228 E. info@blankandcables.com W. www.waltercraven.com

Créa Diffusion, Route de Metz 57, 57580 Remilly, France. T. +39 (0)3 87 646923 F. +39 (0)3 87 647754 E. crea.diffusion@wanadoo.fr

Cristal Quattro, Via di Collandino 200/B, Monteroni d'Arbia 53014, Sienna, Italy. T. +39 (0)577 374231 E. cristal4@ftbcc.it W. www.cristalquattro.it

Dada SpA, Strada Provinciale 31, 20010 Mesero (MI), Italy. E. dada@dadaweb.it W. dadaweb.it

Lorenzo Damiani, Via Segantini 55, Lissone 20035 (MI), Italy. T. +39 (0)2 45 53 42 F. +39 (0)2 45 53 42

Danskina, Pakhuis Amsterdam, Oostelijke Handelskade 15–17, 1019 BL Amsterdam, The Netherlands. T. +31 (0) 20 4198586 F. +31 (0) 20 4198601 E. orga@danskina.nl W. www.danskina.nl

David Design, Stortorget 25, SE-211 34 Malmö, Sweden. T. +46 (0)40 300000 F. +46 (0)40 300050 E. info@david.se W. www.david.se

DeCarloGualla Studio di Architettura, Via Palermo 12, 20121 Milan, Italy. T. +39 (0)2 62 69 05 48 F. +39 (0)2 62 91 02 38 E. decarlo.gualla@tiscalinet.it

de Sede AG, Oberes Zelgli 2, Klingnau 5313, Switzerland. T. +41 56 268 01 11 F. +41 56 268 E. info@desede.ch W. www.desede.ch

Design Gallery Milano, Via Manzoni 46, 20121 Milan, Italy. T. +39 (0)27 989 55 W. www.designgallerymilanol.it

De Vecchi, Via Lombardini 20, 20143 Milan, Italy. T. +39 (0)2832 3365 F. +39 (0)25810 1174 E. info@devecchi.com W. www.devecchi.com

Degre Zero Architecture, 55 Washington Street, Studio 511, Brooklyn NY 11201, USA. T. +1 718 923 1853 F. +1 718 855 9187 W. www.degrezero.com

DMA, 3-12-42-4B Moto Azabu Minato Ku, Tokyo, Japan 106 0046. T. +81 70 5802 1624 F. +81 3 3405 6879

Driade SpA, Via Padana Inferiore 12A, Fossadello di Caorso 29012, (PC), Italy. T. +39 (0)523 81 86 60 F. +39 (0)523 82 23 60

Edizione Straordinaria com Varese, C.so Matteoti 53, Varese 21100, Italy. T. +39 (0)332 232 351 F. +39 (0)332 232 351

Edra SpA, Via Livornese Est 106, Perignano 56030 (PI), Italy. T. +39 (0) 587 61 66 60 F. +39 (0) 587 61 75 00 E. edra@edra.com W. www.edra.com

Elmar Flototto GmbH, Am Ölbach 28, 33334 Gütersloh, Germany. T. +49 (0)5241 9405 0 E. verkauf@elmarfloetotto.de W. www.elmarfloetotto.de

Emmemobili, Via Torino 29, 22063 Cantù (CO), Italy. T. +39 (0) 31 710142 E. emmemobili@emmemobili.it

Eva Denmark A/S, Højnasuij 59, Rødovre DK-2610, Denmark. T. +45 36 73 20 60 F. +45 36 70 74 11 E. mail@evadenmark.com W. www.evadenmark.com

Felicerossi, Via Sempione 17, 21011 Casorate Sempione (VA), Italy. T. +39 (0)331 767131 F. +39 (0)331 768449 W. www.felicerossi.it

Stanislav Fiala and Daniela Polubedovová, D_3A spol. sro, Pístavní 5, 170 00 Prague 7, Czech Republic. T. +420 2 66 71 24 20 E. dastudio@terminal.cz

FiveTwentyOneDesign, 172 E 4th Street, Suite 11A, New York NY 10009, USA. T. +1 212 673 8597. E. info@fivetwentyonedesign.com W. www.fivetwentyonedesign.com

Flos, Via Angelo Faini 2, 25073 Bovezzo (BS), Italy. T. +39 (0)302 711578 F. +39 (0)302 711578

Fresh, 1-39-17-407 Kitazawa Setagaya-Ku, 155-0031 Tokyo, Japan. T. +81 (0)3 5790 0098 F. +81 (0)3 5790 0094 E. fresh@de-code.com

Frighetto Industrie srl, Arzignano Vicenza, Italy. T. +39 (0)444 471717 F. +39 (0)444 451718 E. info@frighetto.it W. www.frighetto.com

Johannes Fuchs Produkt Design, Rückerstrasse 6, Fach 34, D-60314, Frankfurt am Main, Germany. T. +49 (0)69 94410421 E. johannes-fuchs@gmx.net

Fusina srl, Via Rivarotta 11, 36061 Bassano del Grappa (VI), Italy. T. +39 (0)424 590163 E. info@fusina-italy.com W. www.fusina-italy.com

Galerie Kréo, 22 Rue du Chef Delaville, 75013 Paris, France. T. +33 (0)1 53 60 18 42 F. +33 (0)1 53 60 17 58 E. kreogal@wanadoo.fr

The Gallery Moormans, Kzr Karelpln 8b, 6211TC Maastricht, The Netherlands. T. +31 (0)43 3257450

Gärsnäs AB, Box 26, 27203 Gärsnäs, Sweden. T. +46 414 53000 F. +49 414 50616 E. info@garsnas.se W. www.garsnas.se

Ferdi Giardini & Associati, Via Modena 53, 10100 Torino, Italy. T. +39 (0)11 857343 F. +39 (0)11 857343 E. ferdigiardini@virgilio.it

Giovannetti srl, PO Box 1, 51032 Bottegone (PT), Italy. T. +39 (0)573946222 F. +39 (0)573946224 E. info@giovannetticollezioni.it W. www.giovannetticollezioni.it

Stefano Giovannoni, Via Gulli 4, 20147 Milan, Italy. T. +39 (0)2 48703495 E. studio@stefanogiovannoni.it W. www.stefanogiovannoni.it

Glace Control, Rue de la Garenne 37, Pontouvre 16160, France. T. +33 (0)5 45 69 05 15

Glas Platz EK, Auf den Pühlen 5, 51674 Wiehl-Bomig, Germany. T. +49 (0)22 61 7890 0 E. glas-platz@mail.oberberg.de W. www.glas-platz.de

Graves Design Studio, 341 Nassau Street, Princeton NJ 8540, USA. T. +1 609 924 6409 F. +1 609 457 0700 W. www.gravesdesign.com

Happell Interiors and Products, 256 High Street, Glasgow G4 0QT, Scotland. T. +44 (0)141 552 7723 F. +44 (0)141 552 7813 E. info@happell.co.uk

Happy Finish Design, 21 Hillcrest Avenue, Kew, Victoria 3101, Australia. T. +61 3 9817 9780 E. nickrennie@netspace.net.au

Hishinuma Associates Co. Ltd, Tokyo, Japan. T. +81 (0)3 5770 8333 F. +81 (0)3 5770 8334 W. www.yoshikihishinuma.co.jp

Hive, Unit 1.02, Oxo Tower Wharf, Barge House St, London SE1 9PH, UK. T. +44 (0)20 72 61 97 91 E. hive@hivespace.com W. www.hivespace.com

Kate Hume Glass, Kerkstraat 152, 1017GR Amsterdam, The Netherlands. T. +31 (0)20 620 3030 F. +31 (0)20 420 5750 E. kate.hume@euronet.nl

Ideo, White Bear Yard, 144A Clerkenwell Road, London EC1R 5DF, UK. T. +44 (0)207 713 2600 F. +44 (0)207 713 2601 E. sam@ideo.com

IKEA of Sweden AB, Box 702, Tulpanvägen, Älmhult 34381, Sweden. T. +46 476 81363 F. +46 476 15123 W. www.ikea.com

Innovation Randers A/S, 38 Blommevej, Randers 8900, Denmark. T. +45 86 43 82 11 E. mail@inno.dk

Interlübke, Willemstraat 23, 2282 CB, Rijswijk, The Netherlands. T. +31 (0)70 390 33 54 W. www.interlubke.com

ISM Object PTY Ltd, 64-66 Market Street, 3205 South Melbourne, Victoria, Australia. T. +61 3 9645 8881 F. +61 3 9645 8660 E. ismobjects@ismobjects.com.au

ixilab, 6-38 Kazashi-cho, Sakaide, Kagawa 762-0038, Japan. T. +81 (0)9028222310 F. +81 877 440671 E. info@ixilab.com W. www.ixilab.com

Charles O. Job Design & Architecture, Ottikerstrasse 53, CH-8006 Zürich, Switzerland. T. +41 1 361 14 20 F. +41 1 361 14 20 E. jobcharles@gmx.ch

Viktor Jondal, Allen Street 134, Apt. 16, New York, NY 10002, USA. T. +1 917 714 4701 E. viktor@jondal.net

Cordula Kafka Designerin, Schlossallee 45, D-13156 Berlin, Germany. T. +49 30 4766997 F. +49 30 4766997

Kartell SpA, Via delle Industrie 1, 20082 Noviglio (MI), Italy. T. +39 (0)2 900121 F. +39 (0)2 9053316 E. kartell@kartell.it W. www.kartell.it

King-Miranda Associati, Via Forcella 3, Milan 20144, Italy. T. +39 (0)283 949 63 F. +39 (0)283 607 35 E. mail@kingmiranda.com W. www.kingmiranda.com

Knoll, USA. E. A3@knoll.com W. www.knoll.com

Korban/Flaubert, 8/8-10 Burrows Road, St Peters, NSW 2044, Australia. T. +61 2 9557 6136 F. +61 2 9557 6136 E. info@korbanflaubert.com.au W. www.korbanflaubert.com.au

Marc Krusin, Via le Coni Zugna 23, 20144 Milan, Italy. T. +39 (0)348 7040559 E. mkrusin@hotmail.com

Kyo Design, Rue du Sabot 31, 59800 Lille, France. T. +33 (0)3 20 30 82 53 F. +33 (0)3 20 30 02 12 E. contact@kyo-design.com W. www. kyo-design.com

Julia Läufer Gestaltung, Krausnickstrasse 13, D-10115 Berlin, Germany. T. +49 (0)30 69 50 85 85 F. +49 (0)30 69 50 85 86 E. julia.laeufer@diformer.de

Leica Camera AG, Oskar-Barnackstrasse 11, D-35606 Solms, Germany. T. +49 (0)6442 208 215 F. +49 (0)6442 208 333 E. info@leica-camera.com W. www.leica-camera.com

Lexon, 98 ter Boulevard Héloise, Argentueil 95100, France. T. +33 (0)1 39 47 04 00 E. world@lexon-design.com W. www.lexon-design.com

Ligne Roset, Serrieres de Briord, Briord 01470, France. T. +33 (0)4 74 36 17 00 F. +33 (0)4 74 36 16 95

Liv'it srl, Via Macerata 9, 61010 Tavullia (PS), Italy. T. +39 (0)721 202709 E. livit@livit.it W. www.livit.it

K.C. Lo, 31 Finsbury Park Road, London N4 2JY, UK. T. +44 (0)207 359 6791 E. KC@netmatters.co.uk

Lolah, 2265 Royal Windsor Drive, Mississauga, Ontario, Canada. T. +1 800 909 8233 E. info@lolah.com

Lorenz-Kaz, Viale Tunisia 10, Milan, Italy. T. +39 (0)2 29510400 F. +39 (0)2 29510400 E. c.lorenz@iol.it

Macho Products, Inc, 10045 102nd Terrace, Sebastian, FL 32958, USA. W. www.macho.com

Magis srl, Via Magnadola 15, 31045 Motta di Livenza (Treviso), Italy. T. +39 (0)422 768 742 F. +39 (0)422 766 395 E. magisuno@tin.it

Mark Mann, Darmstädter Landstrasse 300, D-60598 Frankfurt/M, Germany. T. +49 69 68 31 64 E. mark-mann@gmx.de

Marburg Tapetenfabrik, J.B. Schaefer GmbH & Co. KG, Bertram-Schaeferstrasse 11, 35274 Kirchhain, Germany. T. +49 (0)6422 81144 F. +49 (0)6422 81222 E. contact@marburg.com W. www.marburg.com

Mathmos Ltd, 20 Old Street, London EC1V 9AP, UK. T. +44 (0)20 7549 2700 F. +44 (0)20 7549 2745 E. mathmos@mathmos.co.uk W. www.mathmos.com

Matsushita Electric Industrial Co. Ltd, 14th Floor National Tower, 2-1-61 Shiromi Chuo-ku, Osaka 540-6214, Japan. T. +81 6 6949 2044

Ingo Maurer GmbH, Kaiserstrasse 47, 80801 München, Germany. T. +49 89 381 606 6 F. +49 89 381 606 20 W. www.ingo-maurer.com

David Mellor Design Ltd, The Round Building, Hathersage, Sheffield S32 1BA, UK. T. +44 (0)1433 650220 E. davidmellor@ukonline.co.uk W. www.davidmellordesign.com

Memphis srl, Via Olivetti 9, Pregnana Milanese 200010, Italy. T. +39 (0)2 93290663 F. +39 (0)2 93591202 E. memphis-milano@tiscalinet.it

Alessandro Mendini, Via Sannio 24, 20137 Milan, Italy. T. +39 (0)2 55185185 E.mendini@energy.it

Micro Mobility Systems D GmbH, Butzensteigleweg 1613, Rosenfeld 72348, Germany. +49 (0)7428 9418 E. info@mircro-mobility.de W. www.micro-mobility.com

Jason Miller, 193 Berry Street, Brooklyn, NY 11211, USA. T. +1 718 218 6636 F. +1 718 218 6636 E. jasonmiller@mindspring.com

Issey Miyake, 5 Place des Vosges, 75005 Paris, France. T. +33 (0)1 4454 5600 W. www.isseymiyake.com

MNO Design, Volmarijnstraat 118, 3021 XW Rotterdam, The Netherlands. T. +31 (0)10 4259696

Molteni EC SpA, Via Rossini 50, 20034 Giussano (Milan), Italy. T. +39 (0)362 3591 E. customer.service@molteni.it W. www.molteni.it

Money For Milan, Mariannenplatz 1, Munich 80538, Germany. T. +49 89 2285262 W. www.moneyformilan.com

Monkey Boys, Hugo Molenaarstraat 39B, 3022 NP, Rotterdam, The Netherlands. T. +31 (0)10 4765 042 E. info@monkeyboys.nl W. www.monkeyboys.nl

moooi©, Jacob Catskade 35, 1052 BT Amsterdam, The Netherlands. T. +31 (0)20 6815051 E. info@moooi.com W. www.moooi.com

Carlo Moretti srl, Fondamenta Manin, 30141 Murano, Italy. T. +39 (0)41 739217 F. +39 (0)41 736282

Mauro Mori, Via G. Compagnoni 3, 20129 Milan, Italy. T. +39 (0)2 70124518 F. +39 (0)2 70124518

Lauren Moriarty, 46 High Street, Rode, Bath BA11 6PB, UK. T. +44 (0)7787 562533 W. www.laurenmoriarty.co.uk

Moroso SpA, Via Nazionale 60, 33010 Cavalicco, Udine, Italy. T. +39 (0)432 577111 E. info@moroso.it

Mutech, 191-1 Anyag-Tdon Manan-ku, Anyang-Ci, Gyongi-Do, Korea. T. +81 31 467 8711 F. +81 31 467 8711 E. yoon@taekwange.com

Nani Marquina, C/Església 46, 08024 Barcelona, Spain. T. +34 93 237 6465 F. +34 93 217 5774 E. info@nanimarquina.com W. www.nanimarquina.com

No Picnic Industrial Designers AB, Ljusslingan 1, 120 31 Stockholm, Sweden. T. +46 55696550. E. ab@no-picnic.se W. www.nopicnic.com

Nube, Via Don L. Meroni 87, 22060 Figino Serenza (CO), Italy. T. +39 (0)31 780 295 F. +39 (0)31 781 958 E. nubeitalia@nubeitalia.com W. www.nubeitalia.com

Ocean North Architecture Design Research, Add. Meritullinkatu 11D, 00170 Helsinki, Finland. T. +358 9 278 3602 W. www.ocean-north.net

OFFECCT Interiör AB, Skövdevägen Box 100, SE-543 21 Tibro, Sweden. T. +46 (0)504 415 00 E. support@offecct.se W. www.offecct.se

Oluce srl, Via Cavour 52, 20098 S. G. Milanese (MI), Italy. T. +39 (0)2 98491435 F. +39 (0)2 98490779 E. info@oluce.com W. www.oluce.com

Palluccoitalia SpA, Via Azzi 36, 31040 Castagnole di Paese (TV), Italy. T. +39 (0)422 238800 W. www.pallucca.com

Panvision, Sintrupvej 35, Bizabrand DK-8220, Denmark. T. +45 86 24 90 24

Pâtisserie Pulicani, Fb St Antoine 172, 75012 Paris, France. T. +33 (0)1 43 72 49 17

Pedrizzetti e Associati, Via Adige 6, Milan 20135 (MI), Italy. T. +39 (0)2 55194324 F. +39 (0)2 5519 4344 E. paolo@pedrizzetti.it W. www.pedrizzetti.it

Phos Design, 11 Hübschstrasse, Karlsruhe 76135, Germany. T. +49 721 849595 F. +49 721 85130 E. info@phos.de W. www.phos.de

Jordi Pigem de Palol, Ronda Ferran Puig 25, Pral. 4A, 17001 Girona, Spain. T. +34 972510835 F. +34 972510835 E. pigem@sct.ictnet.es

PLH Design, Dampfaergevej 10, Copenhagen 2100, Denmark. T. +45 35 43 00 55 E. jn@plh.dk

Poltrona Frau srl, 5577 KM 74 500, 62029 Tolentino, Italy. T. +39 (0)733 909 214 F. +39 (0)733 971600

Radi Designers, Rue de Turenne 89, 75003 Paris, France. T. +33 (0)1 42 71 89 57 E. info@radidesigners.com W. www.radidesigners.com

Rumford Gardner, 225 Newman Avenue, Rumford RI 02916, USA. T. +1 800 234 1957 F. +1 401 431 0882

Denis Santachiara, Alzaia Naviglio Grande 156, 20144 Milan, Italy. T. +39 (0)2 422 1727 F. +39 (0)2 489 53802 E. desanta@tiscalinet.it

Sawaya & Moroni SpA, Via Manzoni 11, Milan 20121, Italy. T. +39 (0)2 8639 5210 E. sawaya-moroni@apm.it W. www.sawayamoroni.com

Sellex SA, Donosti Ibilbidea 84, Astigarraga 20115, Gipuzkoa, Spain. T. +34 943 557011 F. +34 943 557050 E. sellex@adegi.es

Serralunga, Via Serralunga 9, Biella 13900, Italy. T. +39 (0)15 2435711 F. +39 (0)15 31081 E. info@serralunga.com W. www.serralunga.com

Jerszy Seymour Design Workshop, Via Valparaiso 9, Milan 20144, Italy. T. +39 (0)2 43982270 F. +39 (0)2 4391 9507 W. www. jerszyseymour.com

Sharp Corporation, Nagaike-cho 22–22, Abeno-Ku 545–8522, Japan. T. +81 6 6621 3637 E. nisikawa@cdc.osa.sharp.co.jp W. www.sharp.co.jp

Smart Design, 137 Varick Street, New York, NY 10013, USA. T. +1 212 807 8150 F. +1 212 243 8514 W. www.smartdesignusa.com

Elizabeth Paige Smith, 4220 Glenco Avenue, Suite 50, Marina del Rey, California 90292, USA. T. +1 310 306 5010 F. +1 310 306 5035 E. info@epsdesign.com W. www. epsdesign.com

Sony Design Centre Europe, The Heights, Brooklands, Weybridge, Surrey KT13 0KW, UK. T. +44 (0)1932 816163 F. +44 (0)1932 817003 W. www.sony.com

Rainer Spehl, Unit 13c, 20-30 Wilds Rents, London SE1 4QG, UK. T. +44 (0)20 74077302 F. +44 (0)20 74077302 E. rspehl@yahoo.com

Sterling-Miller Designs, 157 Centre Street, 3rd Floor, Brockton MA 02302, USA. T. +1 508 894 6999 F. +1 508 896 6777 E. smdesigns1@juno.com

Christopher Streng Inc, 110 Pine Street, Sheboygan Falls, WI 53085, USA. T. +1 920 467 4026 F. +1 920 467 4668 E. chris@christopherstreng.com W. www.christopherstreng.com

Styling srl, Via Dell'Industria 2, 35010 Borgoricco (PD), Italy. T. +39 (0)49 9318711 F. +39 (0)49 9318700 E. styling@styling.it W. www.styling.it

Swedese Møbler AB, PO Box 156, Formvägen, SE-567 23 Vaggeryd, Sweden. T. +46 393 797 02 E. kristina@swedese.se W. www.swedese.se

Sung Design Limited, Sundial Cottage, Hampton Court Road, East Molesey, Surrey KT8 9DA, UK. T. +44 (0)20 8977 9505 W. www.sungdesign.com

Tagliabue srl, Via Leopardi 1, 22060 Figino Serenza (CO), Italy. T. +39 (0)31 780 604 F. +39 (0)31 781 587 E. info@tagliabuesrl.com W. www.tagliabuesrl.com

Target Stores, USA. T. +1 612 696 3444 W. www.target.com

Tarkett Sommer S.A., 2 rue de l'Egalité, F-92748 Nanterre Cedex, France. T. +33 (0)1 41 20 40 40 F. +33 (0)1 41 21 49 09 W. www.tarkett-sommer.com

Tech International Corp., 1150B E. Hallandale Beach Blvd. Hallandale, FL 33009, USA. T. +1 954 454 1880 W. www.techintlcorp.com

Terra Design, Rua Queiroz Lima 84, 05088-040 São Paulo, Brazil. T. +55 11 3641 1409 F. +55 11 3834 5765 E. terradesign@uol.com.br

Renaud Thiry, Rue de Revilly 73, 75012 Paris, France. T. +33 (0)1 44 73 94 58 E. renaud.thiry@libertyourf.fr

Tonelli srl, Via Della Produzione 33/49, 61025 Montelabbate (PS), Italy. T. +39 (0)721 481172 E. tonelli@tonellidesign.it W. www.tonellidesign.com

Torh Møbler, Wilhelmsgate 3, 0168 Oslo, Norway. T. +47 23 36 76 96 F. +47 23 36 76 75 E. cathrine@torh.no W. www.torh.no

Toshiba Corporation, 1-1 Shibaura, 1 Chome, Minato-ku, 105-8001 Tokyo, Japan. T. +81 3 3457 4022 W. www.toshiba.co.jp

David Trubridge, 44 Margaret Avenue, 4201 Havelock North, New Zealand. T. +64 6877 4684 E. trubridge@clear.net.nz W. www.davidtrubridge.com

Tubes Radiatori, Via Boscalto 32, Resana 31023 (TV), Italy. T. +39 (0)423 480 742 W. www.tubesradiatori.com

Paolo Ulian Industrial Design, Via Silvio Pellico 4, 54100 Massa, Italy. T. +39 (0)585 25 35 73 F. +39 (0)585 25 35 73 E. drugos@iol.it

Umbra Inc, 1705 Broadway, Buffalo, NY 14212, USA. T. +1 800 387 5122 F. +1 416 299 6168 E. suppliers@umbra.com W. www.umbra.com

univorm, Keilestraat 8, 3029 BP Rotterdam, The Netherlands. T. +31 (0)10 4258874 E. info@univorm.nl W. www.univorm.nl

Juan Benavente Valero, Diseñador Industrial, Manuel Martí 5,3, 46021 Valencia, Spain. T. +34 617 31 32 64 F. +34 963 25 11 99 E. juanico@juanico.net W. www.juanico.net

Valigeria Roncato, Via Pioga 91, Campodarsego 35011, Italy. T. +39 (0)49 9290555 E. cristian.roncato@vroncato.it W. www.vroncato.it

Meike van Schijndel, Lange Nieuwstraat 37, 3512 PB Utrecht, The Netherlands. T. +31 (0)30 230 2370 E. m.vanschijndel@worldonline.nl

Versace SpA, Via Manzoni 38, 20121 Milan, Italy. T. +39 (0)2 7609 31 F. +39 (0) 76004122 W. www.versace.com

Vitra AG, Klunenfeldstrasse 22, CH-127 Birsfelden, Switzerland. T +41 (0)61 377 1509 F. +41 (0)61 377 1510 E. info@vitra.com W. www.vitra.com

Vittorio Bonacina, Via Modonnina 12, 1-22040 Lurago d'Erba (CO), Italy. T. +39 (0)31 699800 F. +39 (0)31 699215. E. bonacina@bonacinavittorio.it W. www.bonacinavittorio.it

Von Robinson Industrial Design, VRiD, 13 Ervin Drive, Wappingers Falls, NY12590, USA. T. +1 845 296 1363 E. von@vrid.com W. www.vrid.com

WAAC's, 34/B Hennekijnstraat, 3012 Rotterdam, The Netherlands. T. +31 104126999 F. +31 104128654

waazwiz Ltd., 3-17-8 Aobadai, Meguroku, Tokyo 153-0042, Japan. T. +81 3 5784 0236 F. +81 3 5784 0237 E. waazwiz@cyberoz.net

Norbert Wangen, Seitzstrasse 8/IV, D-80538 München, Germany. T. +49 (0)89 49 00 15 72 E. info@norbert-wangen.com W. www.norbert-wangen.com

Wolf-Gordon, D+D Building, 979 Third Avenue, Suite 413, New York, NY 10022, USA. T. +1 212 319 6800 W. www.wolf-gordon.com

Wovo, 2240W 75th Street, Woodridge, IL 60517, USA. T. +1 800 563 6000. E. farberjohn@aol.com

Wydale Plastics Ltd, Cathole Bridge Road, Crewkerne, Somerset TA18 8RF, UK. T. +44 (0)1460 73212 F. +44 (0)1460 77348

xO, 77179 Servon, France. T. +33 (0)1 60 62 60 60 F. +33 (0)1 60 62 60 62 E. xo@so-design.com

YCAMI SpA, 33 Via Provinciale, 22060 Novedrate (CO), Italy. T. +39 031 789 7311 F. +39 031 789 7330 E. info@ycami.com W. www.ycami.com

Tokujin Yoshioka, 9-1 Daikanyama-cho, Shibuya-ku, 150-0034 Tokyo, Japan. T. +81(0)3 54280830 F. +81(0)3 54280835 E. tokujin8@nifty.com

Zani&Zani, Via del Porto 51/53, 25088 Toscolano (BS), Italy. T. +39 (0)365 641006 F. +39 (0)365 644281 E. zani&zani@tin.it

Zerodisegno, Divisione della Quattrocchio srl, Via Isonzo 51, 15100 Alessandria, Italy. T. +39 (0)131 445361 F. +39 (0)131 68745 E. info@zerodisegno.it W. www.zerodisegno.it

Photographic credits